Exclusive Online Resources for You

As our valued reader, your purchase of this book includes access to exclusive online resources designed to enhance your learning experience. These resources can be downloaded from our website, www.vibrantpublishers.com, and are created to help you apply Java® concepts effectively.

Online resources for this book include the following:

1. Java codes used in the book
2. Additional coding tasks

Why are these online resources valuable:

- **Practical application:** The downloadable codes are provided for easy testing and use.
- **Enhanced learning experience:** The additional coding tasks will provide hands-on experience and help you apply the knowledge learned from the book in practical scenarios.

How to access your online resources:

1. **Visit the website:** Go to www.vibrantpublishers.com
2. **Find your book:** Navigate to the book's product page via the "Shop" menu or by searching for the book title in the search bar.
3. **Request the resources:** Scroll down to the "Request Sample Book/Online Resource" section.
4. **Enter your details:** Enter your preferred email ID and select "Online Resource" as the resource type. Lastly, select "user type" and submit the request.

5. **Check your inbox:** The resources will be delivered directly to your email.

Alternatively, for quick access: simply scan the QR code below to go directly to the product page and request the online resources by filling in the required details.

Happy learning!

SELF-LEARNING MANAGEMENT SERIES

JAVA® ESSENTIALS

VOLUME 2: OBJECT-ORIENTED PROGRAMMING AND BEYOND

Master object-oriented Java programming for professional application development

LAWRENCE G. DECAMORA III, PH.D.

JAVA ESSENTIALS VOLUME 2:
OBJECT-ORIENTED PROGRAMMING AND BEYOND

Published by Vibrant Publishers LLC, USA, www.vibrantpublishers.com

Paperback ISBN 13: 978-1-63651-656-1
Ebook ISBN 13: 978-1-63651-657-8
Hardback ISBN 13: 978-1-63651-658-5

SELF-LEARNING MANAGEMENT SERIES

TITLE	PAPERBACK* ISBN
BUSINESS AND ENTREPRENEURSHIP	
BUSINESS COMMUNICATION ESSENTIALS	9781636511634
BUSINESS ETHICS ESSENTIALS	9781636513324
BUSINESS LAW ESSENTIALS	9781636511702
BUSINESS PLAN ESSENTIALS	9781636511214
BUSINESS STRATEGY ESSENTIALS	9781949395778
ENTREPRENEURSHIP ESSENTIALS	9781636511603
INTERNATIONAL BUSINESS ESSENTIALS	9781636513294
PRINCIPLES OF MANAGEMENT ESSENTIALS	9781636511542
COMPUTER SCIENCE AND TECHNOLOGY	
BLOCKCHAIN ESSENTIALS	9781636513003
CYBERSECURITY ESSENTIALS	9781636514888
MACHINE LEARNING ESSENTIALS	9781636513775
PYTHON ESSENTIALS	9781636512938
DATA SCIENCE FOR BUSINESS	
BUSINESS ANALYTICS ESSENTIALS	9781636514154
BUSINESS INTELLIGENCE ESSENTIALS	9781636513362
DATA ANALYTICS ESSENTIALS	9781636511184
FINANCIAL LITERACY AND ECONOMICS	
COST ACCOUNTING & MANAGEMENT ESSENTIALS	9781636511030
FINANCIAL ACCOUNTING ESSENTIALS	9781636510972
FINANCIAL MANAGEMENT ESSENTIALS	9781636511009
MACROECONOMICS ESSENTIALS	9781636511818
MICROECONOMICS ESSENTIALS	9781636511153
PERSONAL FINANCE ESSENTIALS	9781636511849
PRINCIPLES OF ECONOMICS ESSENTIALS	9781636512334

*Also available in Hardback & Ebook formats

SELF-LEARNING MANAGEMENT SERIES

TITLE	PAPERBACK* ISBN
HR, DIVERSITY, AND ORGANIZATIONAL SUCCESS	
DIVERSITY, EQUITY, AND INCLUSION ESSENTIALS	9781636512976
DIVERSITY IN THE WORKPLACE ESSENTIALS	9781636511122
HR ANALYTICS ESSENTIALS	9781636510347
HUMAN RESOURCE MANAGEMENT ESSENTIALS	9781949395839
ORGANIZATIONAL BEHAVIOR ESSENTIALS	9781636512303
ORGANIZATIONAL DEVELOPMENT ESSENTIALS	9781636511481
LEADERSHIP AND PERSONAL DEVELOPMENT	
DECISION MAKING ESSENTIALS	9781636510026
INCLUSIVE LEADERSHIP ESSENTIALS	9781636514765
INDIA'S ROAD TO TRANSFORMATION: WHY LEADERSHIP MATTERS	9781636512273
LEADERSHIP ESSENTIALS	9781636510316
TIME MANAGEMENT ESSENTIALS	9781636511665
MODERN MARKETING AND SALES	
CONSUMER BEHAVIOR ESSENTIALS	9781636513263
DIGITAL MARKETING ESSENTIALS	9781949395747
MARKETING MANAGEMENT ESSENTIALS	9781636511788
MARKET RESEARCH ESSENTIALS	9781636513744
MODERN ADVERTISING ESSENTIALS	9781636514857
SALES MANAGEMENT ESSENTIALS	9781636510743
SERVICES MARKETING ESSENTIALS	9781636511733
SOCIAL MEDIA MARKETING ESSENTIALS	9781636512181

*Also available in Hardback & Ebook formats

SELF-LEARNING MANAGEMENT SERIES

TITLE	PAPERBACK* ISBN
OPERATIONS MANAGEMENT	
AGILE ESSENTIALS	9781636510057
OPERATIONS & SUPPLY CHAIN MANAGEMENT ESSENTIALS	9781949395242
PRODUCT MANAGEMENT ESSENTIALS	9781636514796
PROJECT MANAGEMENT ESSENTIALS	9781636510712
STAKEHOLDER ENGAGEMENT ESSENTIALS	9781636511511
CURRENT AFFAIRS	
DIGITAL SHOCK	9781636513805

*Also available in Hardback & Ebook formats

To Odette, my wonderful mom, who inspired me to always reach for the stars.

To Michelle, my wife, who inspired me to always do better.

To Danielle and Elisha, our wonderful daughters and our bundle of joy.

To our Lord Jesus Christ, who sustained me and gave us everything that we need according to His riches and Glory.

About the Author

Lawrence G. Decamora III, Ph.D., is an accomplished computer science educator, author, and industry professional with extensive experience in software development and academia. Holding prestigious certifications including Sun Certified Java Programmer (SCJP) and Oracle Certified Professional Java Programmer (OCPJP), he has contributed significantly to the field of information technology.

Dr. Decamora has been a dedicated instructor in computer science, information systems, and information technology for over two decades. He has taught at renowned institutions such as the University of Santo Tomas, De La Salle-College of Saint Benilde, and Mapua Institute of Technology. His academic leadership includes serving as the Acting ITE Program Director and an Academic Head, mentoring countless students and aspiring IT professionals.

In the industry, Dr. Decamora has collaborated with leading organizations like Sun Microsystems Philippines, Phoenix One Knowledge Solutions Inc. , and ActiveLearning Inc., providing Java training and software development solutions. His expertise spans Java programming, enterprise applications, and web technologies.

Beyond teaching, Dr. Decamora is a sought-after speaker and trainer, having conducted seminars and workshops on Java programming, software engineering, and emerging technologies. His passion for knowledge sharing is evident in his commitment to empowering learners through practical insights and real-world applications.

This book is a testament to his dedication to making Java accessible to learners at all levels. Through clear explanations and hands-on examples, Dr. Decamora continues to inspire and guide the next generation of developers.

What Experts Say About This Book!

Java Essentials Volume 2 is a well-crafted text that sets clear learning objectives for each chapter. It leads students through the material with clear, well-coded examples. I support the focus on design and engineering, not "programming."

– John McManus, Ph.D.
Associate Professor of Computer Science
Randolph-Macon College

Having interviewed, technically assessed, mentored, and worked alongside engineers at every level in both start-ups and larger organisations, I've seen how easy it is to drift away from core Java best practices over time. This book is a valuable reference to the fundamentals and a strong guide to staying current with the modern language features, benefiting developers at all stages of their careers.

– Shay Brennan-Kelly
Principal Software Engineer

Finally, a Java book that respects my experience while challenging my complacency. The conversational flow keeps you engaged, the curveball exercises keep you honest, and the depth keeps you coming back. This is technical writing done right.

– Dineshotham Kumar K
Software Engineer
Candescent Technologies

Table of Contents

0 The Bridge to Volume 2: Bridging Fundamentals with Object-Oriented Programming 1

0.1 Looking Back: What You Already Know 2
0.2 A Preview of What's Ahead 5
0.3 Why Object-Oriented Programming? 6
Chapter Summary 9
Multiple Choice Questions 11
Further Reflection 14

1 The Reference Data Types 15

1.1 The Reference Data Types 16
1.2 Assigning References to Variables 20
1.3 The `this` Reference 26
1.4 Local Variables 30
Chapter Summary 40
Multiple Choice Questions 41
Coding Task 45

2 An Introduction to Object-Oriented Concepts 47

2.1 Objects vs Classes 48
2.2 Object Instantiation 52
2.3 Encapsulation - AKA Data Hiding and Implementation 53
2.4 The Constructor 60
2.5 Instance Methods and Static Methods 62
2.6 The Java Source File Layout 63
2.7 Introduction of `modules` 69
2.8 The `JAR` (Java Archive) File 71
2.9 The Manifest File 73
2.10 Java API Documentation 73

Chapter Summary 76
Multiple Choice Questions 78
Coding Task 82

3 Inheritance 87

3.1 Introduction to Inheritance 88
3.2 Classes, Superclasses, and Subclasses 89
3.3 The "is-a" Relationship 91
3.4 Single Inheritance 93
3.5 The Java Access Modifiers 96
3.6 Method Overriding 97
3.7 Rules for Overriding a Method 98
3.8 The super Keyword 100
3.9 Overloading Methods 102
3.10 Rules for Method Overloading 104
3.11 Inheritance and Constructors 109
3.12 Overloading Constructors 112
3.13 The Object Class 115
Chapter Summary 123
Multiple Choice Questions 125
Coding Task 130

4 Polymorphism 135

4.1 Introduction to the Concept of Polymorphism 136
4.2 Virtual Method Invocation: The getDetails() method 140
4.3 Heterogeneous Array 143
4.4 Methods with Polymorphic Arguments 145
4.5 The instanceof Operator 150
4.6 Casting of Objects 153
Chapter Summary 155
Multiple Choice Questions 157
Coding Task 163

5 Other Class Features 169

5.1 Revisiting the `static` Keyword 170
5.2 How do we access/call a static variable or a static method? 172
5.3 The `Math` Class and the `System` Classes 173
5.4 The `static` Imports 174
5.5 The Wrapper Classes 175
5.6 The `final` Keyword 178
5.7 The `record` Keyword 180
5.8 The `enum` Keyword 185
5.9 The `abstract` Keyword 188
5.10 Java Interfaces 195
5.11 The Interface `default` Methods 202
5.12 The use of default methods in implementing multiple interfaces 206
5.13 The Interface `static` Methods 208
5.14 Private Methods in Interfaces (JDK 9) 209
5.15 Sealed Classes (JDK 17+) 211
5.16 The Functional Interface and the Lambda (->) Operator 214
Chapter Summary 217
Multiple Choice Questions 220
Coding Task 225

6 Exceptions and Assertions 233

6.1 The `Exception` Class and the `Error` Class 234
6.2 Java's Approach: The Call Stack Mechanism 236
6.3 General Syntax of an Exception-Handling Block 241
6.4 The `Exception` Hierarchy in Java 249
6.5 Multiple Exceptions in a `catch` Block 251
6.6 The Parameterized `try` Block (Try-With-Resources) 252
6.7 The Handle or Declare Rule 255
6.8 Improved Final or Effectively Final Variable Rethrow (JDK 11+) 256
6.9 Rules on Overriding Methods and Exceptions 258

6.10 Create Your Own Exception 260
6.11 Creating your own unchecked `RuntimeException` 262
6.12 Assertions 264
Chapter Summary 269
Multiple Choice Questions 271
Coding Task 275

7 Building Java Text-Based Applications 277

7.1 I/O Stream Fundamentals 278
7.2 Fundamental Stream Classes 279
7.3 Data Within Streams 279
7.4 I/O Stream Chaining 281
7.5 Java I/O 282
7.6 Reading Input via Keyboard 283
7.7 The `File` class and the `File` Object 289
7.8 Reading Inputs From a File 293
7.9 Writing Inputs to a File 294
Chapter Summary 297
Multiple Choice Questions 299
Coding Task 302

8 Introduction to Java Collection and the Generics Framework 305

8.1 The `Collection` Interface 306
8.2 The `Set` Interface 311
8.3 The `List` Interface 315
8.4 The `Map` Interface 318
8.5 The `Iterator` Interface 322
8.6 An Introduction to the Generics Framework 328
8.7 The Diamond (<>) Operator 331
8.8 Factory Methods for Immutable Collections (JDK 9) 333
8.9 Revisiting Local Variable Type Inference using `var` (JDK 10) 333
8.10 Collectors.toUnmodifiableList/Set/Map (JDK 10) 334

8.11 Sequenced Collections (JDK 21) 336
Chapter Summary 338
Multiple Choice Questions 340
Coding Task 344

9 Building Graphical User Interfaces (GUI) in Java 347

9.1 Introducing the `java.awt` Package 348
9.2 Things to Consider While Building Your GUI Application 351
9.3 Java Layout Managers 357
9.4 Understanding Layout Managers While Creating Your `SimpleCalculator` App 370
Chapter Summary 379
Multiple Choice Questions 380
Coding Task 386

10 Event Handling Techniques 387

10.1 Fundamentals Of Event Handlers 388
10.2 Event Handling Techniques 390
10.3 Building a `SimpleCalculator` App with Event Handlers 404
Chapter Summary 410
Multiple Choice Questions 411
Coding Task 415

11 Building a Swing Desktop App 417

11.1 AWT vs. Swing 418
11.2 Converting Your AWT App to a Swing-Based App 420
11.3 Common Swing Components 431
11.4 Packaging Your Application to a Java ARchive (`JAR`) File 458
Chapter Summary 465
Multiple Choice Questions 466
Coding Task 470

Preface

Java Essentials Volume 2: Object Oriented Programming and Beyond is a natural continuation of the foundational topics introduced in Volume 1. While the first volume focused on building a solid grasp of Java fundamentals—covering syntax, control structures, arrays, strings, and the basic building blocks of Java programs—this second volume shifts the focus to more advanced, structured, and professional programming concepts using **object-oriented programming (OOP)** as the central theme.

This volume is written for learners who are ready to move beyond writing simple programs and begin understanding the true power of Java—**encapsulation, inheritance, polymorphism**, and the wide array of tools that make Java a platform for building scalable, maintainable, and modular software applications.

Starting with a bridge chapter that connects the procedural programming mindset to the object-oriented paradigm, each chapter builds on the previous one, reinforcing understanding through well-structured explanations, sample code, real-world analogies, coding tasks, and review quizzes.

From mastering Java's reference data types to understanding class design, method overloading, inheritance structures, exception handling, and the collection framework, this book is designed to equip students with both theoretical concepts and practical skills. You will also explore **Graphical User Interfaces (GUI)** and the foundations of **event-driven programming**, ultimately culminating in building full-fledged desktop applications using Java Swing.

Whether you're a computer science student, a self-taught programmer, or an instructor looking for a structured resource, Volume 2 serves as a guide to help you think like a Java developer and write code the way professionals do.

I hope that this volume not only helps solidify your Java skills but also inspires confidence in creating applications that are elegant, reusable, and scalable. Let this be your stepping stone toward becoming a proficient Java programmer.

– Dr. Lawrence Decamora
Author & Instructor

Introduction to the Book

Welcome to *Java Essentials: Volume 2*—a comprehensive guide designed to help you master the principles of **object-oriented programming (OOP)** and deepen your understanding of advanced Java topics.

In *Volume 1*, we laid the groundwork by exploring Java's syntax, primitive data types, control structures, arrays, strings, and how to write and run basic Java applications. These topics formed the foundation for thinking like a programmer and writing correct, efficient Java code.

In *Volume 2*, we build on that foundation. This book shifts focus from procedural programming to a more structured and scalable approach using **objects and classes**—the heart of Java programming. Here, you will learn how to design and build real-world applications using OOP concepts such as **encapsulation, inheritance, and polymorphism**, and how to implement them in your own programs.

But we don't stop there.

This volume also introduces you to:

- The **Java API** and how to navigate its documentation effectively
- The concept of **exception handling** to make your programs more robust and error-tolerant
- Working with **Java's Collection Framework** and **Generics** for managing data efficiently
- Creating **text-based** and **graphical user interface (GUI)** applications using AWT and Swing
- Implementing **event-driven programming** and packaging your Java applications for distribution

Each chapter is designed with clarity and progression in mind. You'll encounter:

- **Illustrative code examples** to guide your understanding
- **Hands-on coding tasks** to reinforce concepts through practice
- **Quizzes and chapter summaries** to help you assess your progress
- **Reflection questions and real-world insights** to deepen your critical thinking as a developer

Whether you're a student preparing for exams, a self-taught programmer aiming to level up, or a developer looking to strengthen your Java fundamentals, this book is your next step toward becoming a confident and capable Java programmer.

Let's continue your journey—one object, one method, one application at a time.

Happy coding!

Who Can Benefit From This Book?

Java Essentials Volume 2: Object Oriented Programming and Beyond is written with a wide audience in mind, particularly those who are ready to take the next step in their Java programming journey. If you've completed the basics—whether through *Volume 1* or other introductory courses—this volume is designed to elevate your skills and deepen your understanding of object-oriented and advanced Java concepts.

Here's who will benefit the most from this book:

Students taking object-oriented programming courses

This book aligns with most university-level curricula for second-semester Java programming courses. It introduces key object-oriented concepts like classes, inheritance, polymorphism, interfaces, exception handling, and file I/O in a structured, student-friendly way.

Self-taught programmers

If you've learned Java fundamentals on your own, or with Volume 1, and are now ready to go beyond writing simple scripts or programs, this book will guide you through more sophisticated techniques and design patterns used in real-world Java development.

Instructors and trainers

Educators looking for a well-organized, comprehensive teaching companion will find this volume valuable. The structured format—complete with summaries, quizzes, and coding tasks—makes it easy to build lectures, assignments, and exams around.

Career shifters and bootcamp graduates

If you're shifting into software development or have just completed a coding bootcamp, this book will help you reinforce and expand your skills—especially in object-oriented design, Java Collections, and graphical user interface development.

Junior developers looking to strengthen their foundation

Early-career developers who need to solidify their understanding of core Java and object-oriented programming will find this book an excellent reference and practice guide for everyday development scenarios.

No matter your background, if you're ready to move from beginner to intermediate-level Java programming, this book is designed to support and accelerate your learning journey.

How to Use This Book?

Java Essentials Volume 2 is designed as a step-by-step guide to help learners transition from basic Java knowledge to mastering object-oriented programming and other advanced Java features. Here's how to make the most of it:

1. Follow the chapters in sequence

The chapters in this book are organized progressively. It is strongly recommended to read them in order, as each chapter builds on the concepts covered in the previous ones. For example, understanding reference types and the this keyword in Chapter 1 will make it easier to comprehend inheritance and polymorphism in later chapters.

2. Start with the bridge chapter

Before diving into OOP, Chapter 0 offers a bridge from procedural thinking to object-oriented design. It refreshes key ideas from *Volume 1* and prepares your mindset for writing structured, reusable code using objects and classes.

3. Pay attention to code examples

Each chapter contains carefully selected code snippets that illustrate new concepts. Type these examples on your own, run them in your IDE or terminal, and modify them to deepen your understanding.

4. Try the coding tasks

At the end of most chapters, you'll find **Coding Tasks** that challenge you to apply what you've learned. These exercises range from reinforcing basic concepts to building small functional programs. Use them as practice or mini-projects to test your skills.

5. Use the chapter summaries and quizzes

Every chapter ends with a **Chapter Summary** that recaps the key takeaways, followed by a **Quiz** that reinforces comprehension through multiple-choice or short-answer questions. These tools are especially useful for exam preparation or self-assessment.

6. Explore further with reflections and projects

Some chapters include **reflection questions** or encourage small projects. These are meant to get you thinking critically about how concepts apply to real-world programming scenarios or future software projects.

7. Make use of additional resources

Where applicable, references and screenshots to Java API documentation, additional reading, or helpful tools are provided. Familiarizing yourself with the use of the Java API Documentation will improve your skills in developing Java Apps.

8. Practice consistently

Mastering Java requires consistent practice. Don't just read—code! Use what you learn in each chapter to build small programs, experiment with variations, and try integrating new topics into your past exercises.

This book is more than just a collection of Java topics—it's a roadmap to becoming a confident object-oriented programmer. Take your time, code regularly, and don't be afraid to make mistakes. Every error is an opportunity to grow.

Happy coding!

CHAPTER 0

The Bridge to Volume 2: Bridging Fundamentals with Object-Oriented Programming

Key Learning Objectives

- Review the foundational knowledge from Volume 1 and understand how it connects to object-oriented programming.
- Understand the role of reference data types in object modeling.
- Differentiate between primitive and reference data types through real-world analogies and practical code.
- Recognize the purpose of object-oriented concepts like encapsulation, inheritance, and polymorphism in software development.
- Prepare their learning environment for Volume 2's topics, including multi-class projects, API documentation, and advanced file structures.

0.1 Looking Back: What You Already Know

Before we dive into the more advanced world of Java programming, let's take a moment to look back at everything you've learned so far. In Volume 1, you started your journey by learning how to install Java and set up your development environment. This included installing the JDK (Java Development Kit) and a text editor (i.e., Notepad++). Once everything was set up, you were able to write and run your very first program.

Your first Java program probably looked something like this:

```
// Filename: HelloWorld.java
public class HelloWorld {
  public static void main(String[] args) {
      System.out.println("Hello, world!");
  }
}
```

In this example, you learned about the basic structure of a Java program. The `main` method is the starting point of every Java application, and `System.out.println()` is used to display messages on the screen.

As you wrote more programs, you became familiar with Java's syntax rules. You learned that every statement in Java ends with a semicolon (;), code is grouped into blocks using curly braces `{ }`, and good indentation and spacing help make your code easier to read.

Here's an example with proper indentation:

```
if (score >= 75) {
    System.out.println("You passed!");
} else {
    System.out.println("Try again.");
}
```

Java programs store and process data using primitive data types such as `byte`, `short`, `int`, and `long` for whole numbers, `float` and `double` for decimals, `char` for single characters, and `boolean` for true or false values. You might remember using them like this:

```
int age = 16;
double gpa = 3.75;
char grade = 'A';
boolean isEnrolled = true;
```

Each data type has a specific purpose, and choosing the right one makes your code more efficient and understandable.

To make your programs interactive, you learned how to ask for input using the `Scanner` class. This allowed users to type in answers and data at runtime.

```
// Filename: AskName.java
import java.util.Scanner;

public class AskName {
    public static void main(String[] args) {
        Scanner input = new Scanner(System.in);
        System.out.print("Enter your name: ");
        String name = input.nextLine();
        System.out.println("Hello, " + name + "!");
    }
}
```

This example asks the user for their `name` and then prints a greeting. You used similar programs to read numbers and make simple calculators.

You also explored how to make decisions in your code using `if`, `if-else-if`, and `else` statements. These control structures allowed your programs to choose different paths based on conditions.

```
int score = 85;
if (score >= 90) {
     System.out.println("Grade: A");
} else if (score >= 80) {
     System.out.println("Grade: B");
} else {
     System.out.println("Keep working!");
}
```

You learned about loops, which are used to repeat code. Java provides several types of loops, such as `while`, `for`, and `do-while`.

```
for (int i = 1; i <= 5; i++) {
     System.out.println(i);
}
```

Loops help automate repetitive tasks and make your code shorter and more powerful.

Text in Java is handled by the String class. You learned how to combine strings, check their length, and change their format.

```
String name = "Michelle";
System.out.println("Your name has " +
     name.length() + " letters.");
System.out.println(name.toUpperCase());
```

You also explored more advanced string tools like `StringBuilder` and `StringBuffer` for modifying text efficiently.

Sometimes you need to store a list of values, like test scores or student names. That's where arrays came in.

```
int[] scores = {85, 90, 78, 92};
System.out.println("First score: " + scores[0]);
```

Arrays helped you manage multiple pieces of data using a single variable.

All these concepts—variables, data types, input/output, decisions, loops, strings, and arrays—are the building blocks of every Java program. They prepared you to move on to something even more powerful: object-oriented programming (OOP).

In the next chapters, you'll learn how to use Java to model real-world objects using classes and objects, and how to build larger, smarter, and more organized applications. But don't worry—you'll still use everything you've learned so far!

0.2 A Preview of What's Ahead

This volume begins with a deeper look at something called reference data types. These are different from the primitive data types you've already worked with. Reference data types do not store actual values like numbers or characters. Instead, they store references—or memory addresses—that point (references) to objects. You will learn how to create and use objects, which are the heart of object-oriented programming.

Once you understand objects, you'll move on to some of the most powerful concepts in Java: encapsulation, inheritance, and polymorphism. These are the core principles of object-oriented programming (OOP). Encapsulation means wrapping code and data together and hiding the details from other parts of the program. Inheritance allows one class to inherit traits from another class. Polymorphism lets different objects respond to the same method call in different ways.

You will also learn about constructors—special methods used to set up new objects—as well as the difference between instance methods and static methods. An instance method belongs to a specific object, while a static method belongs to the class itself and doesn't require an object to be used.

Later in this volume, you'll explore abstract classes and interfaces, which allow you to create flexible and reusable code. You will also be introduced to functional programming concepts in Java, including lambda expressions and functional interfaces. These allow you to write shorter and cleaner code.

In addition to learning about code, you'll also discover how Java programs are packaged and shared. You'll learn how to create and use `JAR` (Java Archive) files, how to read Java API documentation, and how to handle exceptions—errors that can happen while your program is running.

By the end of Volume 2, you'll be writing larger, more organized, and more professional programs. You'll have the tools to think like a software developer and build applications that are not only functional but also easy to maintain and understand.

0.3 Why Object-Oriented Programming?

As your programs grow in size and complexity, you'll find that organizing your code becomes more and more important. This is where Object-Oriented Programming, or OOP for short, becomes incredibly useful. OOP is a programming style that focuses on using 'objects' to represent real-world things.

Instead of writing one long block of code that does everything, you use objects to break your program into smaller, manageable parts. Each object has its own properties (also called fields) and behaviors (also called methods). This helps make your program easier to understand, easier to fix when something goes wrong, and easier to expand when you want to add new features.

For example, imagine you're writing a program to store information about students. In a basic program, you might use separate arrays like this:

```
String[] studentNames = {"Matthew","Mark","Luke"};
int[] studentAges = {17, 16, 18};
double[] studentGrades = {89.5, 92.0, 85.75};
```

But what if you have 30 students? Using separate arrays like this can become confusing and hard to manage, especially when your data grows. What if you accidentally mix up the names and grades? It's easy to make mistakes when things aren't grouped together.

Instead, using OOP, you can create a Student class that keeps all the information in one place:

```
// Filename: Student.java
public class Student {
  String name;
  int age;
  double grade;

  public void printDetails() {
        System.out.println(name + " is " + age +
        " years old and has a grade of " + grade);
  }
}
```

Now, instead of keeping student data in separate arrays, you can create objects from the Student class. Each object keeps its own data and knows how to display its own details:

```
Student student1 = new Student();
student1.name = "Matthew";
student1.age = 17;
student1.grade = 89.5;
student1.printDetails();
```

This style of programming not only makes your code easier to read and understand but also helps prevent errors. OOP encourages you to think in terms of real-world objects, which makes programming more intuitive and more fun. As you move forward in this volume, you'll learn how to design and build your own classes, create objects, and use the power of OOP to build better applications.

Chapter Summary

What You Have Learned in Volume 1:

- How to download, install, and configure the Java Development Kit (JDK) for various operating systems.
- The key features and architecture of Java, including JDK vs. JRE, and Java's built-in security mechanisms.
- How to write and debug your first Java program using `System.out.println()` and `JShell`.
- Understanding of Java syntax, including keywords, identifiers, comments, and primitive data types.
- Usage of the `Scanner` class for user input and the `Math` class for mathematical operations.
- Application of Java operators and the correct order of precedence.
- Mastery of control structures: `if`, `switch`, `while`, `do-while`, `for`, as well as `break` and `continue`.
- Proficiency in working with strings using `String`, `StringBuffer`, and `StringBuilder` classes.
- How to create, manipulate, and iterate through one-dimensional and two-dimensional arrays.

Why Volume 2 Matters:

- Introduces Object-Oriented Programming, the core of professional Java development.
- Teaches you how to create user-defined classes and objects, enhancing modularity and reusability.
- Explains critical OOP concepts like encapsulation, inheritance, and polymorphism.
- Learn about exception handling techniques, the Collection Framework, IO Concepts, and building GUI apps that will lead to the deployment of an executable `JAR` file.

- Helps you write cleaner, more organized, and maintainable code.
- Builds on your fundamental knowledge to help you think like a software engineer, not just a programmer.

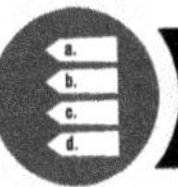

Multiple Choice Questions

1. **Which of the following is a reference data type?**
   a. `int`
   b. `double`
   c. `String`
   d. `boolean`

2. **What is the main characteristic of object-oriented programming?**
   a. Strict memory allocation
   b. Organized around classes and objects
   c. Based on machine instructions
   d. Only uses primitive types

3. **What does encapsulation refer to in OOP?**
   a. Repeating code blocks
   b. Hiding implementation details
   c. Printing variables
   d. Avoiding object creation

4. **Which method is the starting point of any Java program?**
   a. `start()`
   b. `runApp()`
   c. `main()`
   d. `System.out()`

5. **The `Scanner` class is used for:**
   a. Printing to screen
   b. Reading user input
   c. Drawing graphics
   d. Performing mathematical functions

6. **In Java, what does the `new` keyword do?**
   a. Prints variables
   b. Declares a variable
   c. Creates a new object
   d. Deletes an object

7. **What does `String[] args` in the `main` method represent?**
   a. A list of file names
   b. A graphical user interface
   c. Command-line arguments
   d. Random values

8. **What will be the output of `System.out.println(5 + "10")`?**
   a. 15
   b. 510
   c. 10
   d. Compilation error

9. **Which of the following is NOT a Java primitive data type?**
   a. `float`
   b. `boolean`
   c. `char`
   d. `String`

10. **What is the default value of a reference variable if not initialized?**
    a. 0
    b. `false`
    c. `null`
    d. undefined

Multiple Choice Answer Key

1 – c	2 – b	3 – b	4 – c	5 – b
6 – c	7 – c	8 – b	9 – d	10 – c

Further Reflection

1. What were the most important Java programming concepts you learned in Volume 1, and how do you think they will help you in more advanced topics like object-oriented programming?
2. In your own words, what is the difference between a primitive data type and a reference data type? Can you give an example of each?
3. Why is it important to organize code into objects and classes instead of just using variables and methods in a single class?
4. What part of the upcoming topics in Volume 2 are you most interested in (e.g., inheritance, polymorphism, interfaces)? Why does it interest you?
5. How does thinking about real-world objects (like students, cars, or books) help you understand how object-oriented programming works in Java?

CHAPTER 1

The Reference Data Types

Key Learning Objectives

- The reference data types
- Assigning references to variables
- Pass by value
- The `this` reference
- Local variables

In this section, we will delve into the reference data types in Java, which include objects and arrays, allowing variables to reference memory locations rather than storing the actual data. We'll discuss how assigning references to variables works, where variables hold references to objects, enabling multiple variables to point to the same object. Next, we will explain Java's "pass by value" mechanism, where method arguments are passed as copies, meaning changes inside the method do not affect the original variables. Finally, we'll explore the "`this` reference", a keyword used within an instance method or constructor to refer to the current object, helping to resolve naming conflicts and to invoke other constructors or methods.

1.1 The Reference Data Types

Beyond all primitive data types are reference data types. A reference variable contains the address of the object and not the object itself. The object's address is assigned by the Java Virtual Machine (JVM) during the object instantiation stage, commonly known as the creation stage. Hard-coding the object's address in your source code is not possible.

Common examples of reference variables are `Strings`, `Arrays`, and `Objects` lik e `Person` or `Button`.

Example:

```
// Person.java
public class Person {
    public String name = "~noname";
    public int age = 0;

    public Person(String name, int age)      {
        this.name = name;
        this.age = age;
    }

    public String toString() {
        return name + " at age " + age;
    }
}
```

```
1  // TestPerson.java
2  public class TestPerson  {
3      public static void main(String[] args) {
4          Person p1 = new Person("Lawrence", 20);
5      }
6  }
```

Allow me to explain step-by-step what happened in line 4 of the above code snippet.

Figure 1.1 `p1` **initially set to** `null`

```
Person p1 = new Person("Lawrence", 20);
```

p1 → null

`Person p1` created a space in the memory. The reference value will be assigned later. The variable `p1` is called a reference variable because it will later hold a value that is a reference to an object of type `Person`.

Figure 1.2 *`p1`* **initially set to** *`null`* **with initial values of name and age**

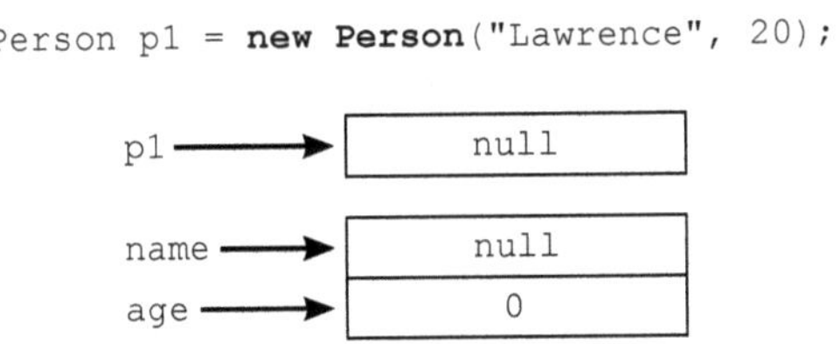

After the space in the heap memory has been reserved for the object, the action now takes place in the stack memory. In Java, stack memory is where local variables and method call information are stored. It operates in a last-in, first-out manner, meaning data is added and removed as methods execute. Heap memory, on the other hand, is used for dynamic memory allocation, where objects and reference data types are created. Based on your class declaration, you have two instance variables: the variable `String name` and the variable `int age`. These variables will also occupy spaces in the stack memory and will be assigned their default values. Table 1.1 shows the default values of each data type.

Table 1.1 Set of default values for variables

Variable	Value
byte	0
short	0
int	0
long	0L
float	0.0F
double	0.0D
char	'\u0000'
boolean	false
All reference types	null

As you can see, since the variable `name` was declared as a `String`, its initial value will be set to `null`. As for your variable `age`, it was declared as an `int`, therefore, an initial value of `0` will be assigned to it by the Virtual Machine.

Figure 1.3 Stack and heap representation of a Java object (Person p1)

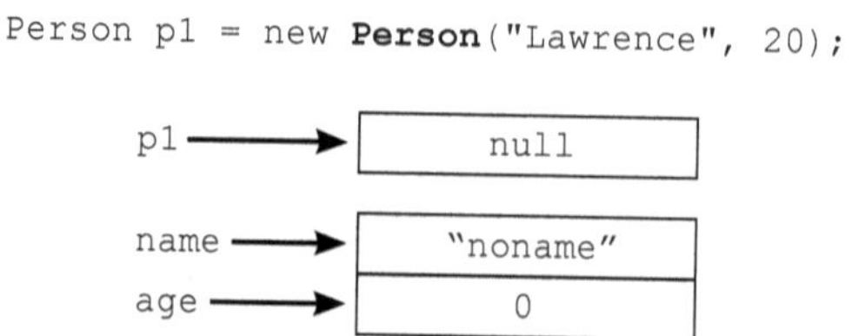

After the initial values are given by the Virtual Machine, the next set of values assigned will be based on the declared values in the class. In the class `Person`, the `String name` will now be assigned a value of `"noname"`, and the `int age` variable will be set to `0`. These values will now be used to overwrite the values that were set by the Virtual Machine.

Figure 1.4 Setting of values for *name* and *age*

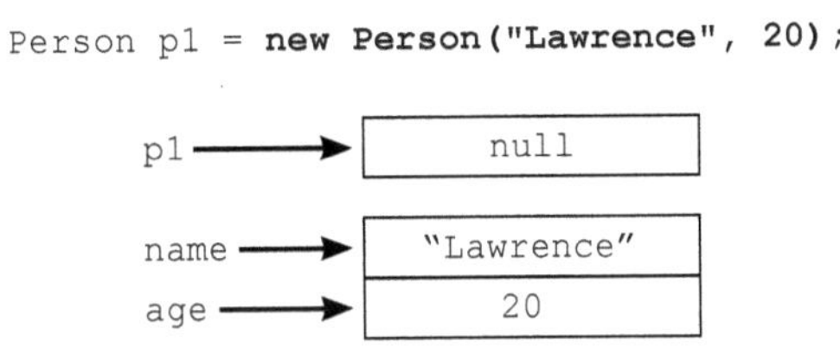

After the class values have overwritten the values set by the Virtual Machine, the next step is to execute the constructor. Executing the constructor means assigning the values of the constructor's parameters to the class's instance variables.

Figure 1.5 Values assigned to the variables after the object is created

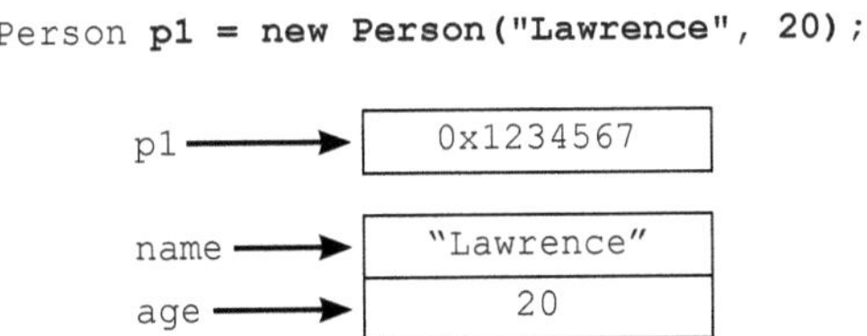

After all parameters of the constructors have been set, all the values of the instance variables, we can safely say that the object has already been created. The last step is to assign the object to its reference variable by assigning its address to it.

1.2 Assigning References to Variables

In lines 1 and 2, both x and y are primitive values. It means that their actual values on the stack memory are 10. While in line 3, you've created a new `Person` object having `"Lawrence"` as the value of its `name` and `20` for its `age`. Since `p1` is a reference variable, the actual value of `p1` is the address where the object resides.

Figure 1.6 **Assigning Values for *x*, *y*, *p1* and *p2***

```
1   int x = 10;
2   int y = x;
3   Person p1 = new Person("Lawrence", 20);
4   Person p2 = p1;
```

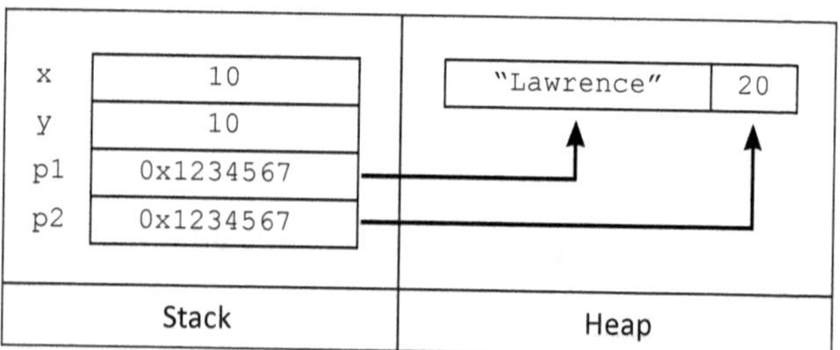

While in line 4, when you say `Person p2 = p1`, what you're actually doing is assigning the value of `p1` to `p2`, which are actually the memory addresses. This means that there are two variables pointing to the same object but having different variable names.

```
5   p2 = new Person("Walter", 30);
```

In line 5, we have created a new object passing a name `"Walter"` with `age` value of `30` and have assigned it to p2. You will see in the diagram below that p2 will now have a new memory address. A new memory address means a new object.

Figure 1.7 Values of *p1* and *p2* after line 5

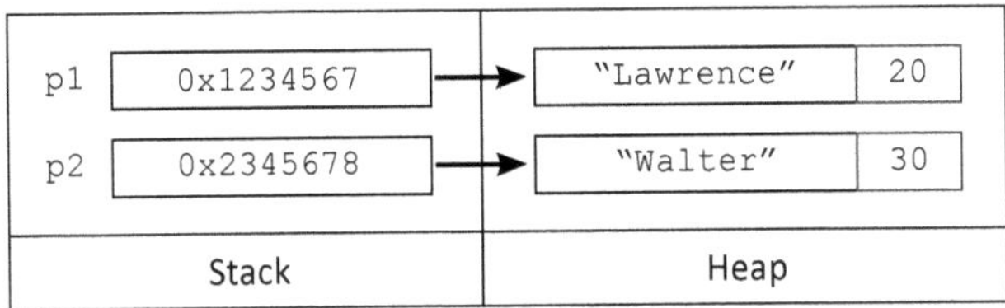

Understanding the difference between primitive variables and reference variables is crucial in Java. Primitive variables store actual values directly in the stack, while reference variables store memory addresses that point to objects in the heap. When you assign one reference variable to another, you're copying the address, not the object itself, meaning both variables point to the same object. However, creating a new object and assigning it to a reference variable results in a new memory address, effectively changing the object that the variable refers to. This distinction is key to mastering object manipulation and memory behavior in Java.

In a single Virtual Machine, everything in Java is Passed by Value.

In a single Virtual Machine, we can pass values to a method, and those values are called arguments. Arguments can be the actual value of a variable, or they can be the address of an object. Here's a sample code that will demonstrate the "Pass by Value" rule.

```
public class PassTest {

    // Methods to change the current values
    public static void changeInt(int value) {
        value = 55;
    }

    public static void changeObjectRef(Person ref) {
        ref = new Person("Walter", 30);
    }

    public static void changeObjectAttr(Person ref) {
        ref.setAge(30);
    }

    public static void main(String[] args) {
        Person per;
        int val;

        // Assign the int
        val = 11;
        // Try to change it
        changeInt(val);
        // What is the current value?
        System.out.println("Int value is: " + val);

        // Assign the person
        per = new Person("Lawrence", 20);
        // Try to change it
```

```
30         changeObjectRef(per);
31         // What is the current value?
32         System.out.println("Person: " + per);
33
34         // Now change the day attribute
35         // through the object reference
36         changeObjectAttr(per);
37         // What is the current value?
38         System.out.println("Person: " + per);
39     }
40 }
```

If this code is executed, we will always start in the main method. In lines 17 and 18, we declared two variables, a primitive variable `val` and a reference variable `per`. If we recall, reference variables contain references to an object, which means they contain the memory address where the object resides. As for primitive types, they contain the actual values of that primitive variable.

In line 21, we initialized `val` to `11` and passed the value of `val` to the method `changeInt(int)`. Since Java uses pass-by-value, the actual value that you're passing to the `changeInt(int)` method is `11` (the actual value). Regardless of what happened inside the method `changeInt(int)`, the actual value of `val` will still be used.

Figure 1.8 **Passing the value of *val* to *changeInt()* method**

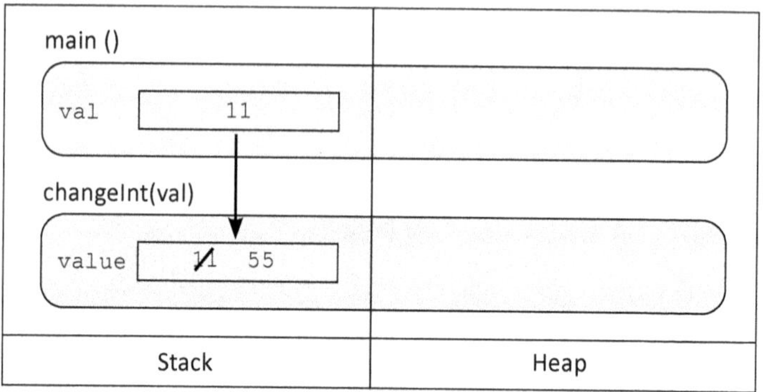

So when you print the output in line 25, the printed output will be:

```
Int value is: 11
```

In line 28, we've created a `new Person` object; thus, the variable `per` will have an address as its value. But in line 30, we passed the value of `per` (that actually contains a memory address) to the method `changeObjectRef(Person ref)` that accepted a `Person` object.

Now, in the method `changeObjectRef(Person ref)` in line 9, we instantiated a `new Person` object, thus letting the argument `ref` point to a new object, thus having a new memory address value to it. In the process, we disregarded its original value that came from the `main` method, which points to the object `"Lawrence" 20`. But that object will not be garbage collected just yet because the variable `per` from the `main` method still has a reference to it. Since the active method is `changeObjectRef(Person ref),` the variable `per` of the `main` method is dormant at the background. Once line 10 is reached, the `changeObjectRef(Person ref)` method ends, and along with that, all local variables are deleted from memory, including the variable `ref`. And since variable `ref` is the only variable that is handling the object `"Walter" 30`, that object will also be deleted through garbage collection.

Figure 1.9 Variable *`ref`* assigned to a new object in the *`changeObjectRef()`* method

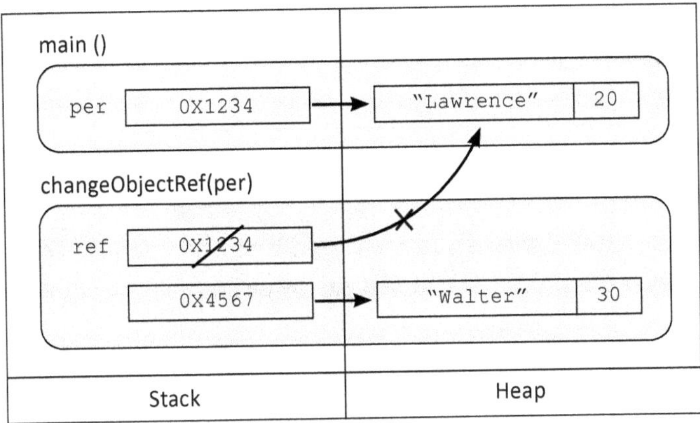

So, when you execute line 32, you will have the output:

```
Person: Lawrence at age 20
```

Lastly, in line 36, you passed the `per` object to the method `changeObjectAttr(Person ref)`, thus passing again the memory address value of `per` to the said method. Inside the method `changeObjectAttr(Person ref)` in line 13, you used the variable `ref` to call a member method `setAge(int)` to change the `age` of the `Person` object that was passed to the `changeObjectAttr(Person ref)` method. While variable `ref` is inside the method `changeObjectAttr(Person ref)`, it actually has access to the object `"Lawrence" 20` that was created in the `main` method because `ref` contains the object's memory address.

Figure 1.10 **The variable `ref` accepts the reference address from the method `main()`**

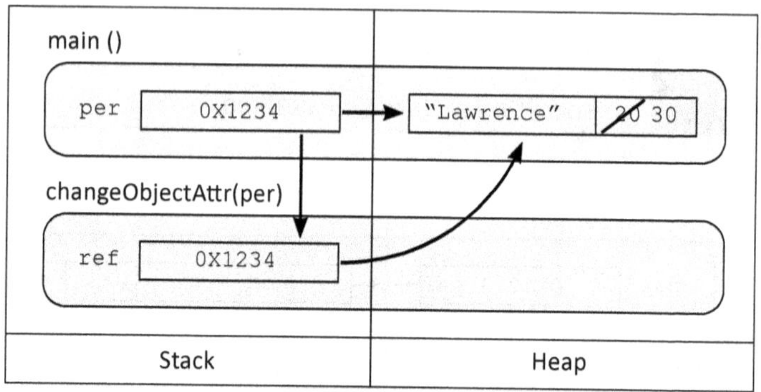

Thus, after the method `changeObjectAttr(Person ref)` executes, the output of line 38 will be:

```
Person: Lawrence at age 30
```

1.3 The `this` Reference

The keyword `this` can be used to:

1. Resolve ambiguity between instance variables and parameters.
2. Pass the current object as a parameter to a method.
3. Pass the current object as a parameter to a constructor.
4. Call another constructor within the same class by using **`this()`** at the first line of the body of another constructor. Consider the given sample code:

```
public class Person {

    private String name = "noname";
```

```
    private int age = 0;

    public Person(String name, int age) {

        this.name = name;
        this.age = age;
    }
    public Person(Person per) {

        this.name = per.name;
        this.age = per.age;
    }
    public void setName(String name) {

        this.name = name;
    }
    public void setAge(int age) {

        this.age = age;
    }
    public String getName() {

        return name;
    }
    public int getAge() {

        return age;
    }
    public Person addAge(int moreAge) {

```

```
            Person newPerson = new Person(this);
            newPerson.age = newPerson.age + moreAge;
            return newPerson;
        }
        public String toString() {
            return name + " at age " + age;
        }
    }
```

```
public class TestPerson {
    public static void main(String args) {
        Person p1 = new Person("Lawrence", 20);
        Person olderPerson = p1.addAge(10);
        System.out.println(olderPerson);
    }
}
```

In class `Person`, we have two constructors, and inside the body of the first constructor that has the signature `public Person(String name, int age)`, we saw that the parameters `String name` and `int age` are identical to our instance variables (attributes); thus we used the keyword **`this`** to resolve the ambiguity inside the constructor's body.

```
this.name = name;
this.age = age;
```

Variable name and age in lines 8 and 9 refer to the local variables name and age declared in the constructor's argument list in line 6. While in the same lines 8 and 9, this.name and this.age refers to the instance variables declared in lines 3 and 4.

Another use of the keyword this is to pass the current object to the method or to the constructor. The current method refers to the method in which the current operation is taking place. If the this keyword is used for the current constructor, it follows the same rule. If the keyword this is used for a constructor, it refers to the constructor where the operation is currently taking place. And while inside the said method or constructor, the keyword this refers to the object that was used to call the said method or constructor.

Consider the method public Person addAge(int moreAge){}, inside the said method, in line 34, you have a constructor call:

```
34      Person newPerson = new Person(this);
```

In this line, the keyword this was passed to the second constructor in line 11.

```
11      public Person(Person per) {
12
13          this.name = per.name;
14          this.age = per.age;
15      }
```

In line 11, the argument Person per will accept the current object that was passed to the said constructor.

By the end of the execution of the program code, you should have been able to identify the flow of the Person object from the `main` method, to the `Person` constructor, back to the `main` method for printing.

1.4 Local Variables

Local variables in Java are variables declared inside a method, a constructor, or a block and are only accessible within that specific scope. These variables are created when the method, constructor, or block is executed and are destroyed once the execution is complete. Since local variables are confined to their scope, they are not accessible outside of the method or block where they are defined. Local variables need to be initialized first before use; they are not automatically initialized.

Here's a code snippet that will cause an error:

```
public void somethingIsWrongHere() {
        int x;
        System.out.println(x);       // error!
}
```

In the method `somethingIsWrongHere()`, we have a local variable x that was declared and printed in the next line. This will cause a compilation error because x was not initialized.

1.4.1 Local Variables in a Method

```
public class LocalVariableExample {
    public void displaySum() {
        int num1 = 10;                  // Local variable
        int num2 = 20;                  // Local variable
        int sum = num1 + num2;          // Local variable
```

```
            System.out.println("Sum: " + sum);
    }

    public static void main(String[] args) {
            LocalVariableExample example =
                new LocalVariableExample();
            example.displaySum();
    }
}
```

Here, num1, num2, and sum are local variables declared inside the displaySum() method. These variables exist only while the displaySum() method is being executed. Once the method completes, these variables are discarded.

1.4.2 Local Variables in a Block

Local variables in a block are variables declared within a specific block of code, such as an if, for, while, or {} block. These variables are accessible only within the block where they are defined, meaning they cannot be used outside of that block. Once the block finishes executing, the local variables within it are destroyed, and their memory is released.

```
public class BlockScopeExample {
  public static void main(String[] args) {
     // This variable exists in
     // the entire main method.
     int outerVariable = 5;

     if (outerVariable > 0) {
        // Local to this if block
```

```
            int innerVariable = 10;
            System.out.println("Inner Variable: "
                                + innerVariable);
        }

        // If used, would cause a compilation error
        // System.out.println(innerVariable);
    }
}
```

1.4.3 Method With Parameters as Local Variables

```
public class MethodWithParameters {
    // 'name' is a parameter and a local variable
    public void greet(String name) {
        System.out.println("Hello, " + name + "!");
    }

    public static void main(String[] args) {
        MethodWithParameters example =
                    new MethodWithParameters();

        // Passing "Lawrence" as an argument
        example.greet("Lawrence");

        // Passing "Michelle" as an argument
        example.greet("Michelle");
    }
}
```

In the `greet()` method, `String name` is a parameter that acts as a local variable. The value of `name` is assigned when the method is called (e.g., `"Lawrence"` or `"Michelle"`). This variable exists only during the execution of the `greet()` method and is destroyed once the method finishes.

The `name` parameter is accessible only within the `greet()` method. But outside of the `greet()` method, `name` cannot be used or referenced.

Method parameters are local variables that exist within the method's scope. They temporarily hold the values passed by the calling code (arguments).

Local variables are essential for temporary data storage within a method or block. They enhance code modularity and efficiency by ensuring variables are accessible only within their defined scope.

1.4.4 Local Variable Type Inference Using `var` (JDK 10)

The keyword `var` is used in place of an explicit type declaration for local variables. The compiler figures out the type by analyzing the expression on the right-hand side of the assignment.

Syntax:

```
var variableName = expression;
```

Here, the `expression` must have a type that the compiler can determine.

Key Points to Remember:

- `var` can only be used for local variables, including inside methods, in loop indexes, and in try-with-resources blocks (which will be discussed in Chapter 6).

- `var` cannot be used for fields (instance or `static` variables), method parameters, or method return types.
- The variable must be initialized at the time of declaration.
- Once inferred, the type is fixed and cannot change later.

Here are simple examples of local variable type inference.

- Inferring an `int`

```
var count = 10;
System.out.println(count + 5);
```

- Inferring a `String`

```
var greeting = "Hello, world!";
System.out.println(greeting.toUpperCase());
```

- Inferring a `double`

```
var pi = 3.14159;
System.out.println(pi * 2);
```

- Loop Indexes - `var` works well in `for` loops:

```
for (var i = 0; i < 5; i++) {
    System.out.println("i = " + i);
}
```

Limitations and Bad Practices:

- Don't use when it hurts readability

```
var x = someMethod(); // What does x hold?
String userName = someMethod(); // Clear and explicit
```

- You cannot use `var` without initialization

```
// Error: cannot infer type without initializer
// var total;
var total = 0; // OK
```

- It cannot be used for method parameters or return types

```
// public void print(var name) {}  // Not allowed
// public var getName() {              // Not allowed
//   return "Alice";
// }
```

Benefits of Using `var`

- Reduces code verbosity
- Improves readability in some cases

In Java 10, the `var` keyword was introduced to simplify local variable declarations by allowing the compiler to infer the variable's type from the value on the right-hand side of the assignment. This feature does not make Java a dynamically typed language; the type is still determined at compile time. When using `var`, it is best to do so where it improves the clarity of your code, rather than simply making it shorter. Additionally, variables declared with `var` must always be initialized at the time of declaration.

Here is an example of the use of the `var` keyword in a complete Java program.

```
public class VarExample {
  public static void main(String[] args) {
     // Using 'var' to declare local variables

     // inferred as String
      var message = "Hello, Java 10 and beyond!";

     // inferred as int
      var number = 42;

      // inferred as double
      var pi = 3.14159;

      // inferred as boolean
      var isJavaFun = true;

      // Printing the inferred variables
      System.out.println(message);
      System.out.println("Number: " + number);
      System.out.println("Pi: " + pi);
      System.out.println("Is Java fun? "
                              + isJavaFun);
      // Using 'var' in a loop
      var fruits =
          new String[] { "Apple", "Banana", "Cherry" };

     // 'fruit' inferred as String
```

```
        for (var fruit : fruits) {
            System.out.println("Fruit: " + fruit);
        }

        // 'var' with an object inferred as Person
        var person = new Person("Lawrence", 25);
        person.introduce();
    }
}

// A simple Person class
class Person {
    private String name;
    private int age;

    Person(String name, int age) {
        this.name = name;
        this.age = age;
    }

    void introduce() {
        System.out.println("Hi, I'm " + name +
                " and I'm " + age + " years old.");
    }
}
```

Key points in the example:

The program demonstrates how to use the `var` keyword in Java, which was introduced in Java 10. The main purpose of `var` is to let the compiler automatically figure out the type of a variable based on the value you assign to it. This is called local variable type inference.

For example:

- `var message = "Hello";` → the compiler knows message is a `String`.
- `var number = 42;` → the compiler knows number is an `int`.
- `var pi = 3.14159;` → the compiler knows `pi` is a `double`.
- `var isJavaFun = true;` → the compiler knows it's a `boolean`.

Once the type is decided, it cannot be changed later. This means `var` does not make Java *"dynamically typed"*. It is still type-safe.

In the program, these values are printed to the console to show how `var` works.

Here's the output of the given program saved as `VarExample.java`

```
Hello, Java 10 and beyond!
Number: 42
Pi: 3.14159
Is Java fun? true
Fruit: Apple
Fruit: Banana
Fruit: Cherry
Hi, I'm Lawrence and I'm 25 years old.
```

Points to Remember

- `var` makes code **shorter and easier to read** while still being type-safe.
- You can use `var` inside methods, loops, and blocks.
- You **cannot** write `var x;` or `var y = null;` because the compiler would not know the type.
- `var` does **not** mean Java is dynamically typed — once the type is inferred, it is fixed.

Chapter Summary

- Primitive data types are basic data types provided by Java. They store simple values directly in memory and are not objects.
- Reference data types include objects and arrays. They store references (memory addresses) to the actual data, rather than the data itself. Examples include instances of classes, interfaces, and arrays.
- User-defined classes are created by the programmer to represent custom data types. These are reference data types because variables of these classes hold references to objects created from the class, rather than the object data itself.
- When you assign one reference variable to another, both variables point to the same object in memory. Modifying the object through one reference will reflect when accessed through the other reference.
- In Java, method arguments are always passed by value. For primitive data types, the actual value is passed. For reference data types, the reference (memory address) is passed, but not the object itself.
- Local Variable Scopes: The scope of a local variable is limited to the block of code (e.g., method or loop) in which it is declared. It cannot be accessed outside this block.
- The `this` reference is a keyword in Java used within an instance method or constructor to refer to the current object. It helps to distinguish between instance variables and parameters with the same name, call other constructors, and return the current object instance.

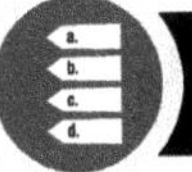

Multiple Choice Questions

1. **Which of the following is considered a Reference Data Type?**
   a. `String`
   b. `int`
   c. `boolean`
   d. `char`

2. **What do you call the classes found in the API Documentation?**
   a. Reference Data Type
   b. Primitive Data Type
   c. Integer Types
   d. Regular Data Types

3. **Given the sample code below:**

```
public class Main{
  private int m = 100;
  public Main(int m) {
    this.m = m;
  }
  public static void main(String args[]) {
    Main myMain = new Main(200);
    System.out.print(myMain.m);
  }
}
```

What will be the output?

   a. 100
   b. 200
   c. 300
   d. 500

4. **Which of the following statements IS NOT TRUE about the `this` keyword?**
   a. You can refer to any member of the current object from within an instance method or a constructor by using this.
   b. From within a constructor, you can also use the `this` keyword to call another constructor in the same class.
   c. The `this` keyword can be used as a reference to a class.
   d. Within an instance method or a constructor, `this` is a reference to the current object — the object whose method or constructor is being called.

5. **Which of the following statements IS NOT TRUE about reference data types?**
   a. Reference data types extend beyond all primitive data types.
   b. A reference variable contains the address of the object and not the object itself.
   c. The object's address is assigned by the operating system during the object instantiation stage, commonly known as the creation stage.
   d. Reference data type values are memory addresses that can be hard-coded in your Java code.

6. **Assuming that a valid class `Person` exists that has a constructor that will accept 2 parameters, a `String name` and an `int age`, the `getName()` method is also provided. Consider the snippet below:**

```
Person p1 = new Person("Lawrence", 20);          // 1

String str = p1;                                 // 2

String name = p1.getName();               // 3
```

Which of the following statements IS NOT TRUE?

a. Line 1 will instantiate the object `p1` from the class `Person`.
b. Line 2 will cause an error.
c. Line 3 will retrieve the name of the person object.
d. All 3 lines of code will compile and run just fine.

7. **Given the method below:**

```
public void doStuff(int x) {

    int y;

}
```

Which of the following is NOT TRUE about the declared local variable in the given method?

a. The local variables `x` and `y` will be deleted from the memory once the method `doStuff()` ends.
b. The local variable `y` will have an initial value of `0`.
c. The local variable `x` will have the value passed to it when the `doStuff()` method is called correctly.
d. The local variable `y` should be initialized first before use.

8. **Which of the following IS NOT a Reference Data Type?**
   a. `float`
   b. `Button`
   c. `Person`
   d. `String`

9. **Which of the following statements IS NOT TRUE?**
   a. A variable has a value related to its data type.
   b. If the variable is a reference type, then you pass in the value, which is the memory address (reference value).
   c. All variables, regardless of the type, and regardless of the scope, will always have a data type and an initial value upon declaration.
   d. When you pass a value to a method, you are passing the variable's value.

10. **Which of the following is considered a Reference Data Type?**
    a. `byte`
    b. `char`
    c. `float`
    d. `Person`

Multiple Choice Answer Key

1 – a	2 – a	3 – b	4 – c	5 – d
6 – d	7 – b	8 – a	9 – c	10 – d

Objective:

Write a Java program that demonstrates your understanding of primitive data types, reference data types, user-defined classes, reference assignments, pass-by-value, local variable scopes, and the this reference.

Requirements:

1. Create a User-Defined Class named `Person`:

Define a class `Person` with the following attributes: `String name` and an `int age`.

Include the following methods in the `Person` class:

- Constructor to accept a `String name` and `int age` value to initialize `name` and `age` attributes.
- `updateAge()` method that accepts a `Person` object and an `int` value to update the `age` of the passed person object.
- `incrementAge()` method that increments the `age` of the current person and returns the updated person object.
- `displayInfo()` method that prints the `name` and `age` of the person.

2. Main Program:

Create a `Person` object and initialize it with a `name` and `age` values as stated in the constructor.

Pass the `Person` object to a method to update its `age` using `updateAge()`.

Use `incrementAge()` to increment the `age` of the `Person` object and assign it to a new reference variable.

Demonstrate the effect of changing the reference variable and local variable scope on the `Person` object.

Use the `this` keyword within the `Person` class to handle method reuse.

CHAPTER 2

An Introduction to Object-Oriented Concepts

Key Learning Objectives

- Distinguish between a class and an object, and explain their relationship in Java.
- Instantiate objects using constructors and understand the role of the default constructor.
- Apply encapsulation to hide implementation details and safeguard object state.
- Differentiate between instance methods and static methods and identify when to use each.
- Describe the structure of a Java source file, including the use of package and import statements.
- Explain the purpose and structure of JAR files, manifest files, and Java API documentation.

In the world of Java programming, understanding object-oriented principles is essential to writing robust, reusable, and maintainable code. This chapter lays the foundation for object-oriented programming (OOP) by introducing the core concepts that define how Java applications are designed and structured. You will learn the differences between classes and objects, how to create and use them, and how encapsulation helps protect internal data. This chapter also covers constructors—including the rules of the default constructor—as well as the distinction between instance and static methods. Finally, we explore how Java source files are organized, how packages and imports work, and how to bundle and document your applications using JAR files, manifest files, and the Java API documentation.

2.1 Objects vs Classes

2.1.1 What Are Objects?

Objects are anything real. They can be something conceptual or something physical. Anything around you right now is an object: the chairs, tables, books, and even your papers. But objects are not just limited to the physical things around us. When you use a desktop application or your favorite word processor, you also see objects in them. Buttons, toolbars, and even the work area or the dialog boxes are objects. Physical objects can be tangible or intangible, and visible, like `Button` and `String`, and invisible, like database connections and network connections.

Objects can also be something conceptual. Conceptual objects are objects that are not seen but exist in a meaningful way. These types of objects are also useful, and they represent a set of data as well. Examples of conceptual objects are errors, exceptions, database connections, network connections, triggered events, etc. We will discuss errors, exceptions, and events in a separate chapter. For now, it is important to understand that these are examples of conceptual objects.

2.1.2 What Are Classes?

A class is the blueprint of an object. It is the declaration of data attributes and methods (also known as behavior) that each object will have every time it is being instantiated. The term "instantiate" is synonymous with the term "to create". So, every time you instantiate an object, you are actually creating an object from a class, and the said object will reside in the heap memory (where Java stores objects and instance variables dynamically at runtime, allowing for flexible management of memory during program execution). You can virtually create an unlimited number of objects from a class. Each object has its own identity. This means that whatever changes take place in one object, the other objects remain unaffected.

In older versions of Java (Java 7 and earlier), objects were always stored in the heap, while class metadata (information about classes, methods, etc.) was kept in a special part of the heap called PermGen (Permanent Generation).

Starting from Java 8, the PermGen area was removed. Objects are still stored in the heap, but class metadata is now stored in Metaspace, which lives in native memory (outside the heap). This change helps avoid memory issues that used to happen with PermGen.

In newer JVMs, the compiler can also optimize very short-lived objects so they don't always end up in the heap—sometimes they may be replaced with variables or stored temporarily on the stack. But from a programmer's perspective, you can think of objects as always living in the heap.

Here's a clear comparison table to summarize Java Memory Structure: Java 7 vs. Java 8+.

Table 2.1 Comparison of Java Memory Structures

Memory Area	Java 7 and Earlier	Java 8 and Later
Heap	Stores all objects	Stores all objects (no change)
Stack	Stores method calls, local variables, and references.	Same as Java 7 (no change)
PermGen	Stores class metadata (class definitions, methods, static variables, etc.)	Removed in Java 8
Metaspace	Not available	Replaced PermGen; stores class metadata in **native memory**
Garbage Collector (GC)	Cleans up heap and PermGen	Cleans up heap and Metaspace (if needed)

Figure 2.1 Diagram that shows the relationship between Classes and Objects

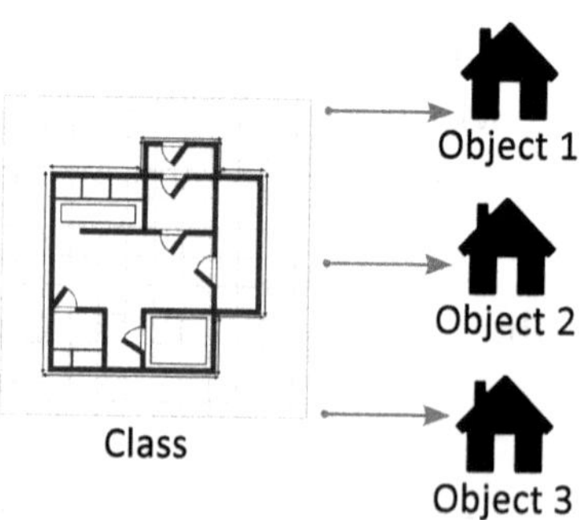

Within this book, we will use several types of UML (Unified Modeling Language) diagrams to graphically represent several figures. In the example in Figure 2.2, we have used a UML Class diagram that will depict the structure of a class. Other types of UML Diagrams are UML Package Diagram, UML State Diagram, etc.

Consider the example in Figure 2.2.

Figure 2.2 UML Diagram of class Person

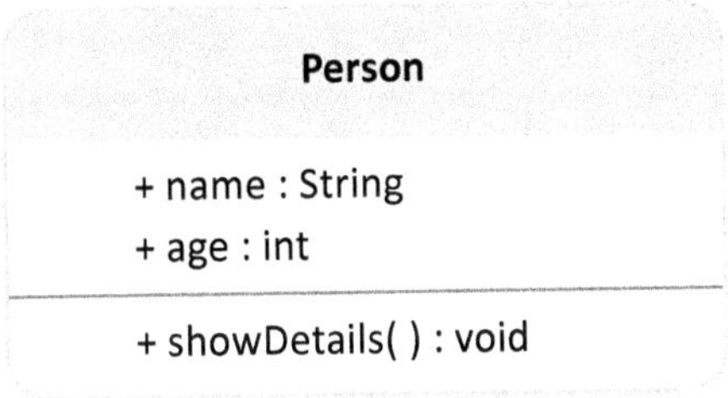

Person
+ name : String + age : int
+ showDetails( ) : void

This UML will have the following Java Class equivalent:

```
public class Person
{
    public String name;
    public int age;
    public void showDetails() {...}
}
```

In the given UML Diagram, the top part (in grey) contains the name of the class, while the second part (white) contains the attributes or variables in the class, and the third part (black) contains the methods in the class. As for their accessibility, the plus (+) means `public`, while in the latter part of this chapter, you will also see a minus (-) symbol that is used to declare something as `private`. Both `public` and `private` access modifiers will be discussed later in a more detailed manner.

2.2 Object Instantiation

Object instantiation in Java is the process of creating an object from a class. You use the `new` keyword along with the class name and parentheses to call its constructor, which sets up the object. This creates a new instance that you can use to access the class's methods and properties.

Let us create two objects.

```
Person p1 = new Person();
Person p2 = new Person();
```

This means that `p1` and `p2` are reference variables that handle two different objects. Since you have two objects now, it means that each object will have its own copy of `name` and `age` attributes, as well as the `showDetails()` method. If you want to access a member (a member is another term for both attributes and methods), you can use the dot (.) notation.

For example, you would like to assign the `name` for the `p1` and `p2` objects, respectively:

```
p1.name = "Lawrence";
p2.name = "Walter";
```

Now, let us assign their respective ages:

```
p1.age = 20;
p2.age = 5;
```

What we've done so far is assign values to the attributes of `p1` and `p2`. Now let us print their values:

```
System.out.println(p1.name + " at age " + p1.age);
System.out.println(p2.name + " at age " + p2.age);
```

This will give us an output of:

```
Lawrence at age 20
Walter at age 5
```

2.3 Encapsulation - AKA Data Hiding and Implementation

Encapsulation helps protect data by keeping it private and only allowing direct access. In Java, you can use the `private` keyword to restrict direct access to a variable, so no one can change it from outside the class. To allow safe access or changes, you use public methods, like getters and setters, which can check or validate values before updating the private variable. This ensures your data stays valid and secure.

What's the problem with this class declaration?

Figure 2.3 UML Diagram of class Person with public variables

Person
+ name : String + age : int
+ showDetails( ) : void

What if we do this?

```
7       Person p1 = new Person();
8       Person p2 = new Person();
9
10      p1.name = "Lawrence";
11      p2.name = "Walter";
12
13      p1.age = 20;
14      p2.age = -5;
```

Will there be any programmatic problems? Will the compiler have a problem with line 14? No. This one will compile and run. The compiler should not have any problems on line 14, because the value `-5` that you're assigning to `p2.age` is a valid integer value. The only problem that this code has is that it represents an unrealistic object.

Java is said to be an object-oriented programming language, which means it has the capability to mimic the behavior of real-life objects. In this case, a real person cannot have a negative value for `age`. In our real world, a person's `age` must be greater than or equal to zero.

So, how are we going to solve this? This is where encapsulation comes in. Encapsulation is an object-oriented feature that hides or protects data from direct access. So, instead of declaring all attributes as `public`, we can declare them as `private` to disable the direct accessibility of the attributes from other classes. Encapsulation helps protect your data by restricting its accessibility.

Figure 2.4 UML diagram of class Person with private variables

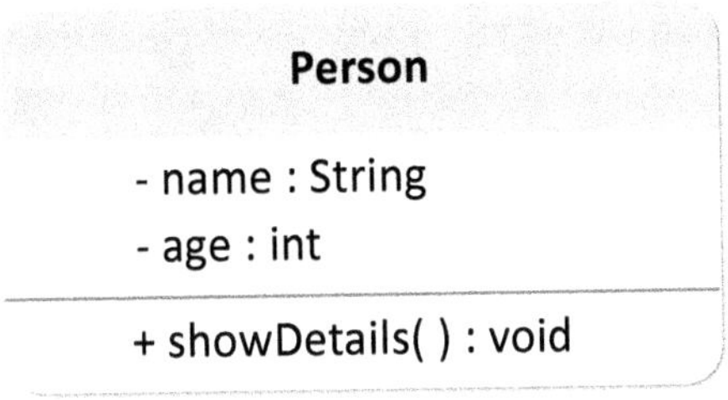

Person
- name : String - age : int
+ showDetails( ) : void

Now, the `name` and the `age` attributes are declared as `private`, which means these attributes can only be accessed within the `class Person` and cannot be accessed outside of this `class`, including the `TestPerson` class.

What if we assign values to `name` and `age` again? We will now have a compilation error because the `name` and `age` attributes were declared as `private`, which means accessing them directly from outside the class is not allowed.

```
p1.name = "Lawrence";          // compilation error
p2.name = "Walter";            // compilation error

p1.age = 20;                   // compilation error
p2.age = -5;                   // compilation error
```

The accessibility of a `private` variable is limited only within the same class where the variable is declared. So, if we declare the `name` and `age` attributes as `private`, how are we supposed to assign values to them and retrieve values from them?

The solution is to provide a method that is publicly accessible and allows access to `private` data; this is where a pair of "setters" and "getters" comes in.

Setters and getters are conventions used to create JavaBeans. They are also used to assign values to the `name` and `age` attributes and can provide verification checks as well.

A Java Bean is just a POJO—Plain Old Java Object—a conventional term for a well-encapsulated class. JavaBeans are not the same as Enterprise Java Beans (EJB). EJBs are the business component of a Java EE Application and are out of our scope.

JavaBeans are reusable software components in Java that follow certain rules, making them easy to use in different programs. One key rule is that a Java Bean must have `private` variables (fields) and provide `public` methods to access or modify these variables, known as *"getters"* and *"setters"*. The *getters* let you read the value of a variable, while the *setters* allow you to change it. This structure ensures that data inside a bean can be managed securely and consistently across different parts of an application.

Here is an example of a Java Bean.

Figure 2.5 UML Diagram of Java Bean class Person

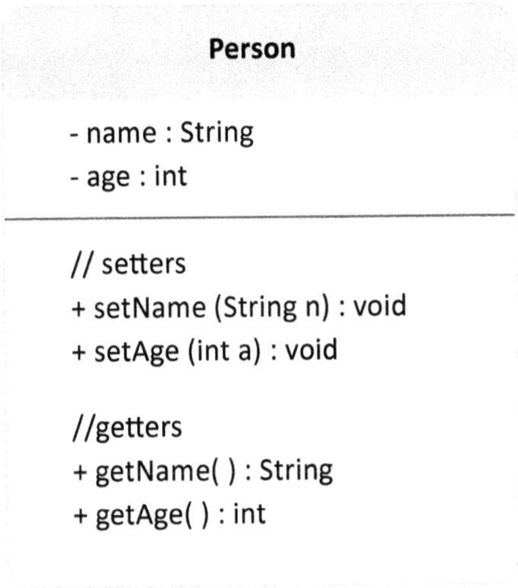

Person

- name : String
- age : int

// setters
+ setName (String n) : void
+ setAge (int a) : void

//getters
+ getName( ) : String
+ getAge( ) : int

Translated into Java code, it will look like this:

```
// Person.java
public class Person {
    private String name;
    private int age;
    // setters
    public void setName(String n) {
        name = n;
    }
    public void setAge(int a) {
        if (a >= 0)
```

```
                  age = a;
            else
                  age = 0;
      }

      // getters
      public String getName() {

            return name;
      }
      public int getAge() {

            return age;
      }

}
```

In this revised `Person` class, you will notice that even if you try to directly assign a negative value for the person's age using `p2.age = -5`, it will not be possible; a compilation error will occur. All values passed directly to the `private` variables from a different class will not be allowed. By providing `public` setters, all values that will be passed to the `private` variables will now be forced to be passed through the `public` setters. With this, you can now provide validation checks inside the `public` setters if needed. All values passed can either be checked or assigned directly.

Let's see our revised `TestPerson` class.

```
// TestPerson.java
public class TestPerson {

```

```
4       public static void main(String[] args) {
5
6           Person p1 = new Person();
7           Person p2 = new Person();
8
9           // assign names
10          p1.setName("Lawrence");
11          p2.setName("Walter");
12
13          // assign age
14          p1.setAge(30);
15          p2.setAge(-5);
16
17          System.out.println(p1.getName() + " is "
                                + p1.getAge() + " years old.");
18          System.out.println(p2.getName() + " is "
                                + p2.getAge() + " years old.");
19      }
20  }
21
```

In lines 10 and 11, as well as lines 14 and 15, values intended for `name` and `age` are now forced to pass through the `public` setters. Therefore, the `Person` class can perform validation checks on these values.

In retrieving the values of `name` and `age`, we will now be forced to call `public` getters.

```
System.out.println(p1.getName()+ " is "
                    + p1.getAge()+ " years old.");
System.out.println(p2.getName()+ " is "
                    + p2.getAge()+ " years old.");
```

Public getters are methods intended to return a value. Once called in lines 17 and 18, a `String` value and an integer value will be returned, respectively. The `p1.getName()` will return a `String` value, and `p1.getAge()` will return an integer value that will be printed via the `System.out.println()` method. The returned values of the getters can either be printed to the screen or stored in a variable. Once compiled and executed, it will have this output:

```
Lawrence at age 30
Walter at age 0
```

2.4 The Constructor

The constructor is a method-like construct that has the same name as your class name. Constructors do not have any return type, not even void.

Here's an example:

```
1 public class Person {
2
3     public Person (String n, int a) {
4
5           name = n;
6           age = a;
7     }
8 }
```

Lines 3 to 7 are your constructor body. To create a `Person` object, you call your constructor this way.

```
Person p1 = new Person("Lawrence", 20);
Person p2 = new Person("Walter", 18);
Person p3 = new Person();     //compilation error
```

The Default Constructor's Rule:

"There is always at least one constructor for each class; if the writer does not provide a constructor, Java will provide the default, no-argument, empty-bodied constructor." For example:

```
public class MyClass {

    // nothing here

}
```

Does this class contain a constructor? The answer is YES. That is because of the default constructor's rule. This is the reason we can instantiate an object from this class. The given class will actually look like this after compilation:

```
1  public class MyClass {
2
3          // this is the default constructor
4          public MyClass() {
5
6              // nothing here
7          }
8  }
```

Lines 4 to 7 will be automatically inserted after compilation, but will only be present inside the class file and not inside the Java source file. The default constructor is provided so that even if the writer does not provide a constructor, we can still create an object from a class such as this one:

```
MyClass mc = new MyClass();
```

Creating an object from this class is actually a call to the default constructor provided by Java.

You are also allowed to have multiple constructors in a class. Having multiple constructors will allow you to have different ways to instantiate (or create) an object. Take, for example, the `java.lang.Character` class. It only contains a single constructor, which means you can create a `Character` object only one way. If you look at the `java.lang.String` class, it contains several constructors, which means that there are two different ways of creating a `String` object.

2.5 Instance Methods and Static Methods

Methods are used to create a subroutine that is called any number of times from the main method or from another method. A method can execute with or without passing values to them; these values are called arguments or sometimes called parameter lists.

Here is the syntax for declaring a method in Java:

```
<modifiers> <return_type> methodName(<parameter_list>)
{
    // method body
}
```

The **`common modifiers`** can be any of the following:

```
- public/private/protected
- static/abstract/final
```

The **`return_type`** can be any data type or even `void`.

The `methodName` can be any valid identifier

The **parameter_list** can be any variable and value pair. While calling an instance method, an object must first be created. An instance method is called through the created object. Given the sample instance method call:

```
p1.setName("Lawrence");
System.out.println(p1.getName());
int nameLength = p1.getName().length();
```

Instance variables and methods belong to an object and are used through that object. To access them, you first create an object of the class using the `new` keyword. Then, you can use the object's name followed by a dot `('.')` to call its methods or access its variables.

Whereas a static method is called through the class name.

```
String strNum = "100";
int num = Integer.parseInt(strNum);
```

The `Integer` class is found under the `java.lang` package and has a `static` method `parseInt()` that can accept a `String` integer format that will return an `int` value. Since the `parseInt()` method is `static`, we call the method through the name of the class, **Integer.parseInt**`(strNum)`.

2.6 The Java Source File Layout

Just remember **P-I-C**, which stands for **P**ackage–**I**mport-**C**lasses. All these parts of the Java Source file are optional. You can have the following number of statements per part:

```
package --> 0 or 1 statement
import  --> 0 to many
classes --> 0 to many
```

A Java source code can have multiple classes, but it is only allowed to have one `public class`. Once a `public class` is declared, then the filename of your Java source code must match the said `public class`.

2.6.1 Java `package` Statement

Packages in Java refer to subdirectories. A package may contain classes or sub-packages. Take the diagram in Figure 2.6 as an example.

Figure 2.6 Sample Package Diagram

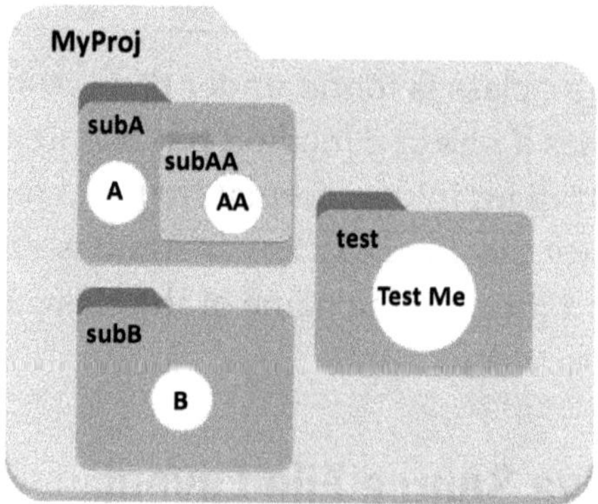

We have `MyProj` as a package, and within `MyProj`, we have three sub-packages, namely: `subA`, `subB`, and `test`. And within `subA`, we have `subAA` as a sub-package.

A package may also contain classes inside it. The package `subA` contains `A.class`, `subAA` contains `AA.class`, package `subB` contains `B.class`, and package `test` contains `TestMe.class`, which serves as our test class.

In using packages, we also need to be aware of the following rules.

1. The basic syntax of a package is this:

```
package <top_level_pkg_name>[.<sub_pkg_name>]*;
```

Example:

```
package subA.subAA;
```

2. The package declaration can only be placed at the beginning of a Java Source File

3. Only one (1) package declaration per source file is allowed.

4. In the absence of the package statement, the generated class file will be placed in the default package (the present working directory).

5. Package statements must be separated by dots (.) to make the subdirectory representation platform-independent.

2.6.2 The `import` Statement

Import statements are used to tell the compiler where to find the needed classes that your application will use. There are two packages that do not need any `import` statements:

1. the present working directory and;

2. the `java.lang.*;` package.

Classes inside the above-mentioned packages do not require any `import` statement declarations. However, classes found on other packages must be declared using an `import` statement. Here's an example:

```
1 import java.util.ArrayList;
2 import java.io.*;
3 import subA.subAA.AA;
```

Line 1 states that the `ArrayList` class can be accessed inside the `java.util` package, while Line 2 states that all classes under the `java.io` package can be used. The **asterisk (*)** signifies **all classes,** while line 3 states that this application is using the class `AA`, which is located under `subA.subAA` package.

Putting it all together, here's an example. This is the `TestMe.java` source file.

```
package test;

import subA.*;
import subA.subAA.AA;
//import subB.B;

public class TestMe {

    public static void main(String[] args) {

        A a = new A();
        AA aa = new AA();

        // fully-qualified name
        subB.B b = new subB.B();

        a.printMe();
```

```
18              aa.printMe();
19              b.printMe();
20          }
21  }
```

Line 1 states that once this source file compiles, the `TestMe.class` that will be generated will be placed under the `test` subdirectory.

Lines 3 and 4 state that classes `A.java` and `AA.java` are needed inside this application and are located under the declared packages.

Line 5 is commented out. If this is the case, you need to use the fully qualified name in line 15 so that you can still get access to the `B.java` source file. In effect, the `import` statements found on lines 3, 4, and 5 tell the compiler where to find the necessary classes so that you don't always use the fully-qualified name of a class. Instead of always using `subB.B b = new subB.B();` you just have to use:

```
import subB.B;
....
B b = new B();
```

Figure 2.7 shows a preview of the developer's directory structure.

Figure 2.7 Complete Project Directory Structure

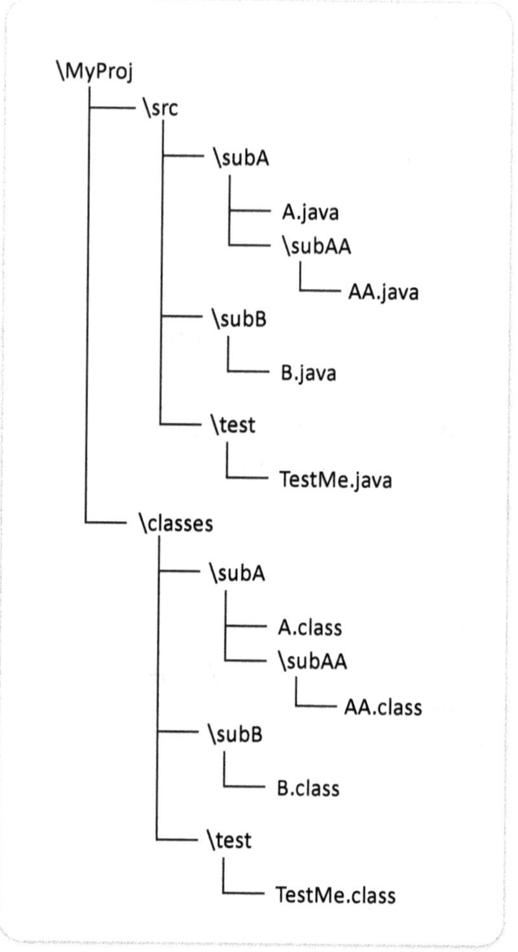

Compiling the test class `TestMe.java` under the `src` folder will also compile the rest of the `*.java` files within the `src` folder. This is because, inside the test class, there are dependencies with other Java source files.

To compile on Windows, use the following command:

```
MyProj\src> javac -d ..\classes test\TestMe.java
```

If this command is successful, the generated class files should be under the `classes` folder.

To run the sample application on Windows, you can type in:

```
MyProj\src> java -cp ..\classes test.TestMe
```

Here, the `-cp ..\classes` option sets the classpath to the classes directory so that the JVM knows where to find the compiled `TestMe.class` file. The fully qualified class name `test.TestMe` is used because the file is inside the `test` package.

For modern IDEs (Integrated Development Environment), you don't have to worry about executing these commands;, they're all automatically managed by your IDE.

2.7 Introduction of `modules`

When working with `modules` (introduced in Java 9), the idea is to group related packages together and describe them using a special file called `module-info.java`. For example, if your project has a `module` named `myapp`, you would create a `module-info.java` file inside the `src\myapp` folder. This file declares the module name and can also specify which packages it exports and which other modules it requires. To compile a module, you use a command like `javac -d ..\out --module-source-path src src\myapp\module-info.java src\myapp\test\TestMe.java`.

Here, the `--module-source-path` tells the compiler where to find the source modules, and `-d ..\out` saves the compiled classes in the `out` folder, keeping them separate from the source.

To run the program, you use a command such as `java --module-path ..\out -m myapp/test.TestMe`.

The `--module-path` tells the JVM where to find compiled modules, `-m` specifies the `module` and the `class` with the `main` method to run.

Table 2.2 Packages vs. Modules in Java

Feature	Packages	Modules
What it is	A way to group related classes into folders.	A higher-level way to group packages into a single unit.
File needed	No special file required (just a folder with `.java` files).	Requires a `module-info.java` file that describes the module.
Organization	Keeps classes organized by functionality (e.g., `test`, `utils`).	Keeps whole packages organized and defines clear boundaries.
Visibility	Classes in a package can be imported by any other code.	A module can **export** only the packages it wants to share and **hide** others.
Dependencies	No direct way to state which packages are required.	`module-info.java` can declare which other modules are required.
Introduced in	Available since the very beginning of Java.	Introduced in **Java 9** as part of Project Jigsaw.
Example	`package test;` → groups classes inside `test`.	`module myapp { exports test; }` → defines a module named `myapp` and exports `test` package.

Packages are used to organize classes into folders so that related code is grouped together in a neat and structured way. Starting from Java 9, modules were introduced as a bigger building block — they allow you to group several packages into one unit and also control which parts of your code can be accessed by other modules. In short, packages help organize classes, while modules help organize packages.

2.8 The JAR (Java Archive) File

A complex application is normally composed of a set of class files and not just one class file. When these class files need to be delivered to our client, we need to put them inside one file called the JAR File or the Java Archive File. This JAR file represents a group of classes that work together.

Consider the given directory structure earlier. After a successful compilation, the generated class files are placed under the classes folder.

Figure 2.8 Partial Project Directory Structure, the classes folder

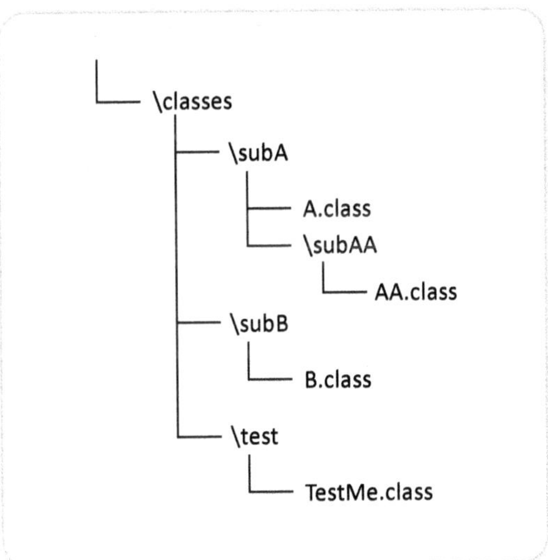

These class files, which are under the classes folder, can now be archived. Archiving a set of classes can be done by the jar utility that comes with the JDK installation.

There are two ways to create a Jar File.

1. Without a manifest file

to create:

```
jar cvf <JarFile.jar> <Files>
```

i.e.

```
jar cvf MyJar1.jar *
```

to run:

```
java -cp <JarFile.jar> <fullyQualifiedName>
```

i.e.

```
java -cp MyJar1.jar test.TestMe
```

2. With a manifest file, also known as the executable Jar File.

to create:

```
jar cvfm <JarFile.jar> <manifestFile> <Files>
```

i.e.

```
jar cvfm MyJar2.jar manif.txt *
```

to run:

```
java -jar <JarFile.jar>
```

i.e.

```
java -jar MyJar2.jar
```

2.9 The Manifest File

A manifest file is a configuration file that is used by an archive file. All JAR files contain a manifest file. If the creator of the JAR file did not include their own manifest file, there will always be the default MANIFEST.MF file that contains the default configuration. By putting your own version of your manifest file, you can customize its behavior and override the default manifest file inside your JAR file.

The manifest file has four parts.

```
<Main-Class tag/s><separator><value><ENTERKEY>
```

A manifest file looks like this:

Manifest-Version: 1.0 <ENTERKEY>

```
Main-Class: test.TestMe <ENTERKEY>
```

Be sure to add a next line character after the last line so that during runtime, the Java Virtual Machine can read the said line input. In the absence of the ENTER KEY character in your manifest file, the JVM will treat this as an invalid manifest file because it will not be able to read its contents.

For most IDEs, the manifest file is also automatically created. You just have to edit it and include the fully qualified name of the test class in the Main-Class tag. This is required if you want the JAR file to be an executable (runnable) JAR file.

2.10 Java API Documentation

As discussed in *Java Essentials Volume 1*, the Java Runtime Environment (JRE) is composed of two main components:

1. The Java Virtual Machine (JVM) and the;

2. Application Program Interface (API)

To know the classes inside your API, you could download the API documentation on the Oracle® website.

> You can also access the API documentation online via the URL:
>
> http://docs.oracle.com/javase/<javaSEversion>/docs/api/index.html
>
> Example:
>
> https://docs.oracle.com/javase/23/docs/api/index.html
>
> Note: Links are also provided in the Online Resources provided along with the book.

Figure 2.9 Java 23 API Documentation Page

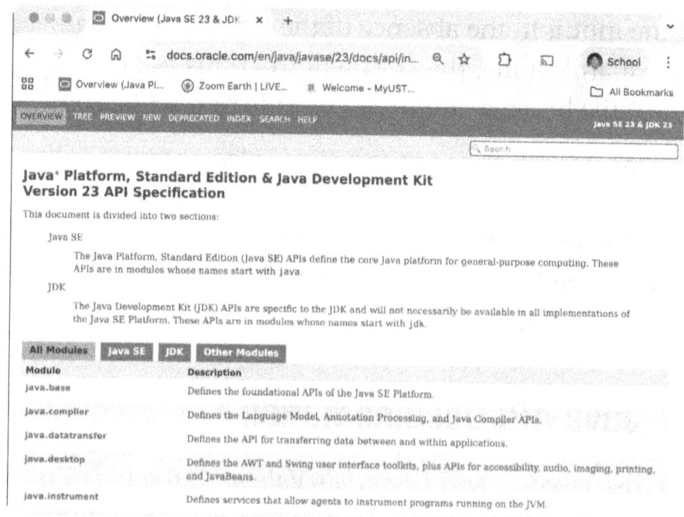

Source: https://docs.oracle.com/en/java/javase/23/docs/api/index.html

Try to use the `Search` bar on the upper right part of the screen and key in the name of the class, i.e., `String`. You should see the write-ups about the said class, the constructors, and the methods of the said class, and other details about it.

Chapter Summary

In this chapter, we talked about the following:

- Objects can be physical or conceptual entities, representing real-world items or abstract concepts like database connections and network links. In software, objects include elements like buttons and toolbars, as well as errors and events. Classes, on the other hand, are blueprints for objects, defining their data attributes and behaviors. When a class is instantiated, it creates an object stored in heap memory, allowing flexible memory management. Multiple objects can be created from a single class, each with its own identity, unaffected by changes in others. UML diagrams, such as class, package, and state diagrams, are used to visually represent these structures.
- A constructor is a special method in Java that shares the same name as the class and has no return type. Every class has at least one constructor; if none is defined by the programmer, Java automatically provides a default, no-argument constructor. This allows objects to be created even if the class appears empty. For instance, when an object of '`MyClass`' is instantiated, Java uses the default constructor to enable this, ensuring smooth object creation.
- Methods in Java are reusable blocks of code that can be called from the `main` method or other methods, with or without passing values (called arguments or parameters). Methods can have various access modifiers (like `public`, `private`, or `protected`) and can be instance methods (requiring an object to be called) or static methods (called directly through the class name). For example, `p1.setName("Lawrence")` is an instance method called on an object, while `Integer.`

`parseInt(strNum)` is a static method accessed via the class `Integer`.

- A Java Source File Layout can have either one package statement or none at all. For import statements, you can have multiple import statements or none at all. For classes, you can have multiple non-public class declarations or have a single public class with several non-public classes or none at all. Packages, import statements, and class declarations are all optional. If they are present, their sequence must start with packages, followed by import statements, and then the list of classes.
- A complex application is normally composed of a set of class files and not just one class file. When these class files need to be delivered to our client, we need to put them inside one file called the `JAR` File or the Java Archive File. This `JAR` file represents a group of classes that work together.

Multiple Choice Questions

1. **What refers to a method-like construct that has the same name as your class name and is used to instantiate (or create) an object from the class?**
   - a. Classes
   - b. Methods
   - c. Constructors
   - d. Garbage Collector

2. **Which of the following IS TRUE about static methods?**
   - a. You can access the static methods through the name of the class.
   - b. You have to instantiate an object first before you can access a static method.
   - c. You can only access the static methods through the name of the object.
   - d. You CANNOT overload static methods.

3. **In the Java code snippet:**

```
int myMethod(){
int myNumber = 12;
return myNumber;
}
```

What is the return type of the method?

- a. `int`
- b. `void`
- c. `myMethod`
- d. none

4. **Which of the following statements is NOT TRUE?**
   a. If the method's return type is an `int`, the values that you pass to a method's parameter MUST also be an int.
   b. A method may return an `int` and can also have a different parameter type.
   c. If a method DOES NOT return any value, the return type should be of type `void`.
   d. If the method's return type is an `int`, the method must include a `return` statement inside it returning an `int` value.

5. **Three of the choices below are attributes for a dog object; which one is a method?**
   a. `age`
   b. `bark()`
   c. color
   d. breed

6. **You have written a class called `Animal` that contains the following:**

```
public void eat() {}

public void breath() {}

public void sleep() {}
```

   What do you call the given contents of the class?
   a. comments
   b. attributes
   c. constructors
   d. methods

7. **Given the following class declaration:**

```
public class Person {
     Person(String name){}
     Person(String name, int age)      {}
}
```

Which of the following will cause an error in creating an instance of the `Person` **class?**

a. `Person p = new Person("Lawrence", 20);`
b. `Person p = new Person();`
c. `Person p = new Person("Percy");`
d. `Person p = new Person("Jerry", 25);`

8. **There are three objects and a class in the options below; which of the options is a** `class`**?**

a. iPhone X
b. Nokia 3310
c. Samsung Galaxy
d. Smartphone

9. **Given the line of code that creates a** `Student` **object:**

```
Student aStudent = new Student();
```

Which of the following operators is used in creating a `Student` **object?**

a. `Student`
b. `=`
c. `aStudent`
d. `new`

10. What does the Java API documentation provide?

a. An Integrated Development Environment (IDE) for writing Java code

b. Detailed information about the syntax and usage of Java keywords

c. Descriptions and usage of classes and methods provided by the Java platform

d. Tutorials and code samples for learning Java

Multiple Choice Answer Key

1 – c	2 – a	3 – a	4 – a	5 – b
6 – d	7 – b	8 – d	9 – d	10 – c

Coding Task 1: Complete the Student Class

Given the partial code for the class Student that contains 3 attributes: `name`, `studentNumber`, and `course`.

You are to provide the setter and getter methods, as well as the body of the constructor that has a missing body.

You are also required to provide an implementation of the method `showStudentInformation()` so that once it is called, it should display the `name`, `studentNumber`, and `course` of the Student object.

```
// This is the code for Student.java
// complete the code.
public class Student {

    private String name;
    private int studentNumber;
    private String course;

    // this is constructor #1 -
    // no need to add anything here
    public Student()
    {}

    // this is constructor #2 -
    // you have to provide the body of this constructor
    public Student(String name, int studentNumber,
                   String course) {

        // complete the body of this constructor
    }

    // put your setters here
    // put your getters here
```

```
    public void showStudentInformation() {

        // this method should display the name,
        // studentNumber and the course of the student
        // object.
    }
}


// This is the code for TestStudent.java -
// complete the code.

public class TestStudent {

    public static void main(String args[]) {

        Student s1 = new Student("Lawrence", 1111,
                           "BS Computer Science");
        Student s2 = new Student();
        Student s3 = new Student();

        // set the properties of Student s2,
        // assign a name, studentNumber and course

        // set the properties of Student s3,
        // assign a name, studentNumber and course

        // print the identity of Student s1
        s1.showStudentInformation();

        // print the identity of Student s2 by
        // calling the getter methods
        // print the identity of Student s3 by
        // calling the showStudentInformation() method
    }
}
```

Coding Task 2: Create and Use a Simple Class

Instructions:

1. Create a class named Book with the following private fields:
   - String title
   - String author
   - int yearPublished
2. Add a constructor that initializes all three fields.
3. Provide public getter and setter methods for each field.
4. Add an instance method named `displayInfo()` that prints the book's details in the format:

   `"Title: <title>, Author: <author>, Yr: <yrPublished>"`
5. In the `main` method of a `class` named `LibraryApp`, instantiate at least two `Book` objects, set or modify their data using setters, and call `displayInfo()` for each object.

Coding Task 3: Use Static Methods and Explore Java File Structure

Objective:

Practice creating static methods, using the `package` and `import` statements, and building a basic project structure.

Instructions:

1. Create a package named `utilities`.
2. Inside the `utilities` package, define a `class` called `MathHelper` with the following:
   - A `static` method `square(int number)` that returns the square of the number.

- A `static` method `isEven(int number)` that returns `true` if the number is even, `false` otherwise.

3. Create a second package named `mainapp`.
4. Inside `mainapp`, create a class `MainProgram` with a `main` method that:
   - Imports and uses the `MathHelper` class.
   - Calls both methods from `MathHelper` using sample data (e.g., 12 and 7).
   - Prints the results.

CHAPTER 3

Inheritance

Key Learning Objectives

- Define and implement inheritance in Java using extends.
- Distinguish between superclasses, subclasses, and the "is-a" relationship.
- Override methods safely following the rules of overriding.
- Overload methods and constructors while adhering to overloading rules.
- Utilize super to refer to superclass members and constructors.
- Explain the role of `Object` and override its methods `toString()`, `equals()`, and `hashCode()`.

This chapter covers the fundamentals of inheritance in Java, from understanding the "is-a" relationship and single inheritance to accessing members of a parent class with the `super` keyword. It focuses on the essential principles of how subclasses are created based on existing classes, allowing for code reuse and clear hierarchies. You'll learn how to override and overload methods, handle constructors in subclasses, and appreciate the role of `Object`—the root of all Java classes—through its `toString()`, `equals()`, and `hashCode()` methods. We will also explore the significance of access control and method behavior, particularly when extending functionality, and how these techniques come together to form a strong, reusable, and well-structured code base.

3.1 Introduction to Inheritance

Inheritance is another fundamental concept of object-oriented programming. The idea behind inheritance is that you can create new classes that are built on existing ones. When you inherit from an existing class, you reuse (or inherit) its methods and variables, and you add new methods and variables to adapt your new class to new situations. This technique is essential in most object-oriented programming languages, particularly in Java.

If you are coming from a procedure-oriented language like C, Visual Basic, or COBOL, you will want to read this chapter carefully. For experienced C++ programmers or those coming from another object-oriented language like `Smalltalk`, this chapter will seem largely familiar. There are many differences between how inheritance is implemented in Java and how it is done in C++ or in other object-oriented languages.

3.2 Classes, Superclasses, and Subclasses

Let's look at the `Employee` class.

Figure 3.1 The *Employee* class

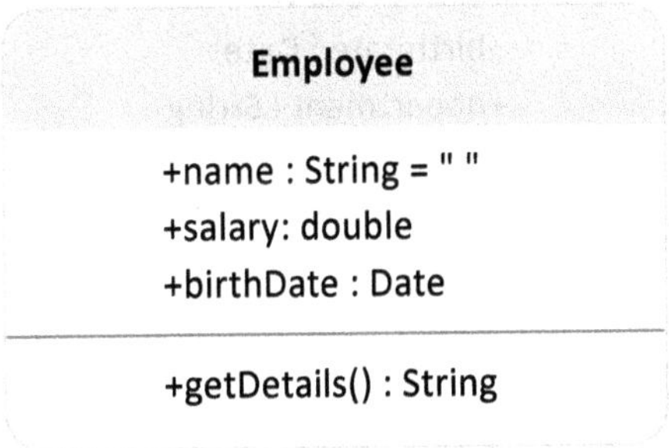

Employee
+name : String = " " +salary: double +birthDate : Date
+getDetails() : String

Suppose you are currently an employee object. As an `Employee`, you have the following attributes: `name`, `salary`, `birthDate`, and a method that will display all your information, let's call it `getDetails()`.

```
public class Employee {
 public String name = "";
 public double salary;
 public Date birthDate;
 public String getDetails() {...}
}
```

`Managers`, in most cases, are also `Employees`. Both `Employees` and `Managers` have their respective `names`, `salaries`, and `birthDates`. On top of these attributes, `Managers` also have their assigned `departments` to manage and run.

Figure 3.2 The *Manager* class

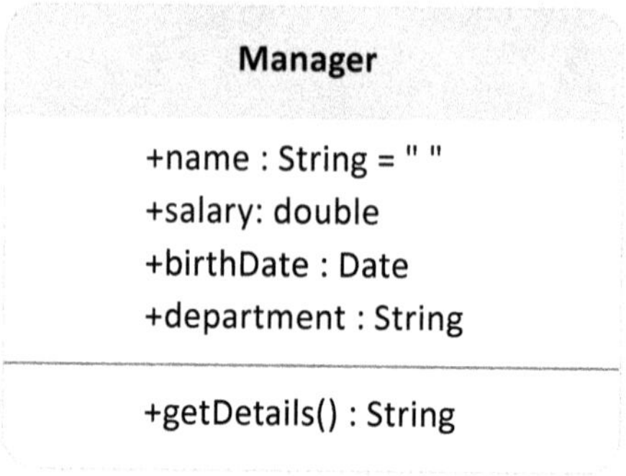

Manager
+name : String = " " +salary: double +birthDate : Date +department : String
+getDetails() : String

```
public class Manager {
      public String name = "";
      public double salary;
      public Date birthDate;
      public String department;
      public String getDetails() {...}
}
```

This is the kind of situation that cries out for inheritance. Inheritance is a way to reuse an existing class by extending it to be its superclass (aka parent class). In our example, the `Manager` class is inheriting the `Employee` class, which means that whatever the `Employee` class has, the `Manager` class will also have.

Figure 3.3 The *`Employee`* and *`Manager`* class

Employee
+name : String = " " +salary: double +birthDate : Date
+getDetails() : String

Manager
+name : String = " " +salary: double +birthDate : Date +department : String
+getDetails() : String

3.3 The `"is-a"` Relationship

In all inheritance cases, there is an obvious `"is-a"` relationship between the parent class and the child class. In Java, an `"is-a"` relationship means that a subclass is a specific type of its superclass. It implies a hierarchical connection where the subclass inherits characteristics and behaviors from the superclass.

For Java technology, we will use the terms superclass and subclass. The `Manager` and `Employee` classes are examples of an `"is-a"` relationship. Every manager *is an* employee. This `"is-a"` relationship is a way to show inheritance.

Here is how you define a `Manager` class that inherits from the `Employee` class. You use the Java keyword `extends` to denote inheritance.

This is the syntax to apply inheritance:

```
class <SubClass> extends <SuperClass> {
 // added fields, constructors, and methods
}
```

To apply the given syntax in the `Employee` and `Manager` classes, we can write the inheritance relationship this way:

```
class Manager extends Employee {
 // added fields, constructors, and methods
}
```

This shows that the `Manager` class is the subclass and the `Employee` class is the superclass, and if we apply the "is-a" relationship to both classes, we can say that a `Manager` "is an" `Employee`.

Figure 3.4 The *Manager* class extending the *Employee* class

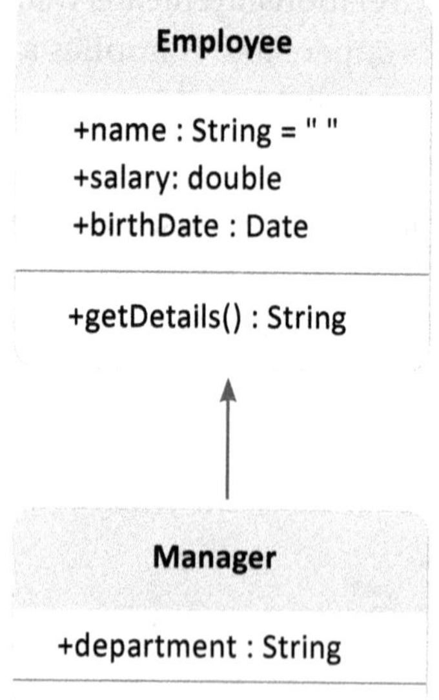

```
// Employee.java
public class Employee {
      public String name = "";
      public double salary;
      public Date birthDate;
      public String getDetails() {...}
}
// Manager.java
public class Manager extends Employee {
      public String department;
}
```

The "is-a" relationship is not just limited to the immediate subclass; it can go further down the hierarchy.

3.4 Single Inheritance

In the Java Programming Language, we can only implement **single inheritance**, which means you can only inherit from a single superclass (parent class). Although it does not limit the superclass to have a single subclass, it can have multiple subclasses.

Notice the UML diagram of the classes in Figure 3.5. Each class only has one superclass. In this UML, you will also notice that a superclass has multiple subclasses, like the class `Employee` has several direct subclasses like `Manager`, `Engineer`, and `Secretary`, while the class `Director` is a subclass of the class `Manager`, which is also a subclass of the `Employee` class. To apply the "is-a" relationship, we can therefore say that the `Director` "is-a" `Manager`. It is also true to say that the `Director` "is an" `Employee`. The `Director` may not be an immediate subclass of the

`Employee` class, but it still belongs to the same hierarchy, so a Director "is an" Employee as well.

Figure 3.5 Single inheritance hierarchy example

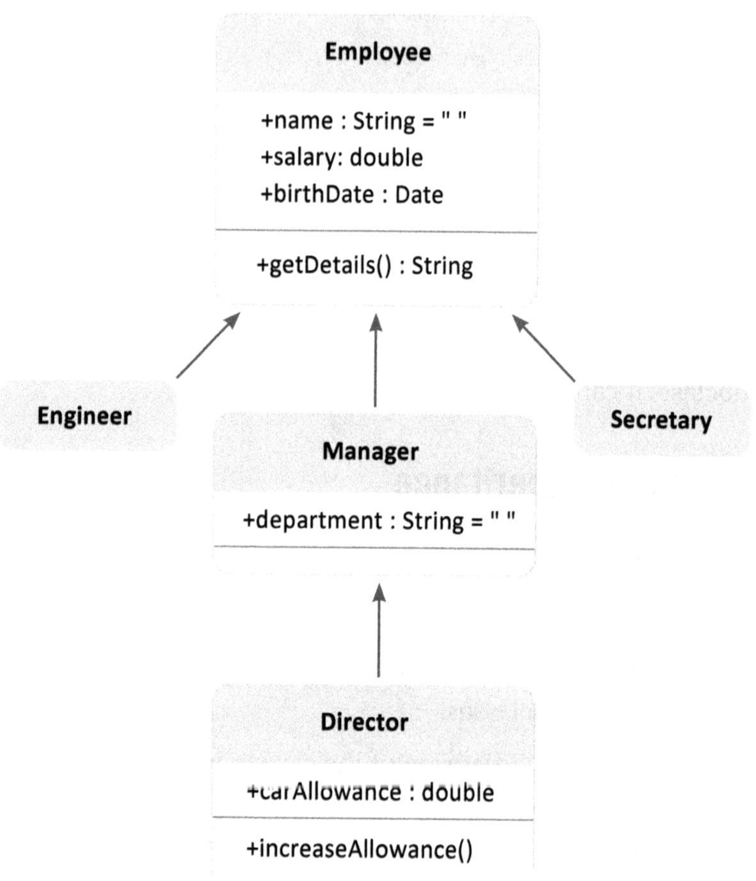

In C++ and other object-oriented programming languages, multiple inheritance is allowed, but not in Java. This begs the question: Is Java inferior to C++? The answer is "NO". Multiple inheritance has several drawbacks. Take Figure 3.6 for example.

Figure 3.6 Multiple inheritance concept

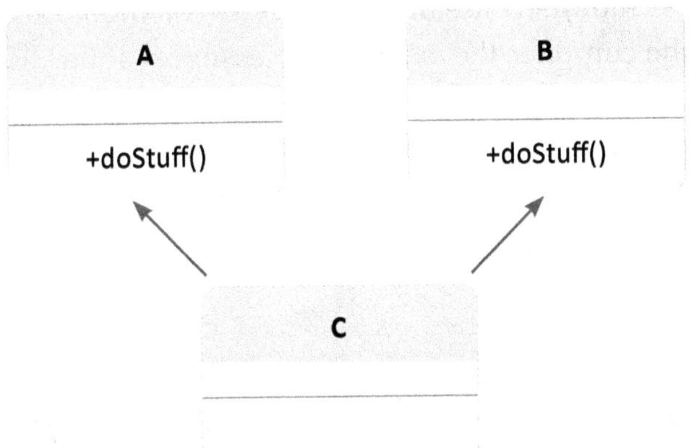

Let us say that `class A` and `class B` are both parent classes, and they have a common method called `doStuff()` that contains two different method implementations. Another class named `class C` is a child class of both `class A` and `class B`.

Which `doStuff()` method will be inherited by `class C`? The `doStuff()` method is present in both parent classes. This causes the problem as to which `doStuff()` method should `class C` inherit, i.e., should it inherit the `doStuff()` method in `class A` or the one from `class B`.

This is a potential problem that can occur with multiple inheritance. This is NOT allowed in Java, but in other programming languages like C++ and Python, this is allowed. Multiple inheritance can lead to errors or ambiguity in design.

In Java, we can simulate multiple inheritance by using `interfaces`. Interfaces and their advantages will be discussed in Chapter 5. Simply put, Java requires more organization to ensure it is more robust when it comes to object-oriented concepts such as inheritance.

3.5 The Java Access Modifiers

Access modifiers are used for classes, methods, and fields. It tells the compiler the extent of accessibility it has. We have four access modifiers in Java. Table 3.1 presents a summary of these modifiers and their respective accessibility.

Table 3.1 Java access modifiers

Access Modifiers	Same Class	Same Package	Subclass	Anywhere
private	✓	✗	✗	✗
<default>	✓	✓	✗	✗
protected	✓	✓	✓	✗
public	✓	✓	✓	✓

Once an inner class, a method, or a variable is declared as `private`, its accessibility is limited only within the class where it was declared. Top-level classes are not allowed to be declared `private`. If none of the access modifiers are used, it implicitly means that you are using the `<default>` access modifier, which limits its accessibility inside the class where it was declared and within the same package-private (sub-directory).

The `protected` access modifier, on the other hand, allows the accessibility to be limited only within the same class, same package, or within the same subclasses, even if they are in different packages. A construct that uses the `protected` access modifier cannot be accessed outside the same package or by classes that are not a subclass.

Lastly, the `public` access modifier can be accessed anywhere. It has universal accessibility. Top-level classes can only have a `<default>` or `public` accessibility. Inner classes, variables, and methods can have any of the aforementioned access modifiers.

3.6 Method Overriding

There are certain scenarios when you inherit a method from a superclass, but you realize that the method you have inherited is not quite enough to perform the duties required by the subclass. To resolve this dilemma, you can override the method you have inherited.

```
// Employee.java
public class Employee {
  public String name = "";
  public double salary;
  public Date birthDate;
  public String getDetails() {
        return "Name: " + name +
              "\nSalary: " + salary;
  }
}
```

```
// Manager.java
public class Manager extends Employee {
  public String department;
  @Override
  public String getDetails() {
        return "Name: " + name +
              "\nSalary: " + salary +
              "\nDepartment: " + department;
  }
}
```

In this example, the `getDetails()` method of the `Manager` class overrides the `getDetails()` method of the `Employee` class. It is because the `Manager's` `getDetails()` method is supposed to give out more information than the inherited `getDetails()` method of the `Employee` class.

3.7 Rules for Overriding a Method

In the source code given, we can say that the `getDetails()` method of the `Manager` class is the overriding method. In overriding a method, a couple of rules need to be observed.

1. The overriding method must have the same name.
   - For example, in the given example, both the `Employee` class and the `Manager` class have the `getDetails()` method.
2. The overriding method must have the same argument list.
   - In this example, both the `getDetails()` method of the `Employee` and the `Manager` class have an empty argument list.
3. The return type of the overriding method can be the same.
   - In this example, both the return type of the `getDetails()` method of the `Employee` and the `Manager` classes are `String` objects.
4. The overriding method must not be less accessible.

- In the `Employee` class, the `getDetails()` method was declared as `public`; this means that the `getDetails()` method of the subclass must also be declared `public`.

5. The return type of the overriding method can be a subclass of the return type of the method of the superclass.

    - In this example, the return type of the `getDetails()` method is of type `String`, and there are no subclasses of the `String` class.

Assuming the `getDetails()` method returns a non-final object, say a non-`String` object, then it is possible for the overriding `getDetails()` method to have a different return type. Here is another alternative version of method overriding to emphasize rule #5:

```
// Employee.java
public class Employee {
    . . .
    public Employee getDetails() {
        // returns Employee object...
    }
}

// Manager.java
public class Manager extends Employee {
    . . .
    @Override
    public Manager getDetails() {
        // returns a Manager object
    }
}
```

A subclass can modify the implementation of the inherited method with a different method implementation than the superclass's method, but it must have the same name, same argument list, and return type. But for Java 5.0 and newer versions the return type of the overriding method can be a subclass of the return type of the superclass's method. Also, the access modifier of the overriding method must not be less accessible.

3.8 The `super` Keyword

In most cases, overriding a method can produce redundancies, which is a poor programming practice. Having redundancies in your code slows down both the development and debugging time by increasing code issues during renaming or refactoring of code, and it also slows down your runtime.

To minimize the problem of redundancy, we can use the `super` keyword. The `super` keyword is used in a class to refer to its superclass's member data – both attributes and methods. The behavior does not have to be in the immediate superclass; it can be further up in the hierarchy.

Here are sample codes for both the `Employee` and `Manager` classes with overriding methods that have redundancies in the `getDetails()` method.

```
// Employee.java
public class Employee {
    public String name = "";
    public double salary;
    public Date birthDate;
    public String getDetails() {
        return "Name: " + name +
```

```
                    "\nSalary: " + salary;
    }
}
// Manager.java
public class Manager extends Employee {
    public String department;
    @Override
    public String getDetails() {
        return "Name: " + name +         // redundant! Bad.
                "\nSalary: " + salary +  // redundant! Bad.
                "\nDepartment: " + department;
    }
}
```

Here's a revised version of the `Employee` and `Manager` code:

```
// Employee.java
public class Employee {
    public String name = "";
    public double salary;
    public Date birthDate;
    public String getDetails() {
        return "Name: " + name +
                "\nSalary: " + salary;
    }
}

// Manager.java
public class Manager extends Employee {
    public String department;
    @Override
```

```
4       public String getDetails() {
5           return super.getDetails() + // this is better.
6               "\nDepartment: " + department;
7       }
8   }
```

In line 5 of the `Manager` class, we used `super.getDetails()` to call the `getDetails()` method of the `Employee` class instead of copying lines 6 and 7, which allows us to get rid of the code redundancy.

3.9 Overloading Methods

In most object-oriented programming languages, we can declare multiple methods with the same name and different argument list and with different method implementations. This language feature is known as method overloading.

Figure 3.7 **Screenshot of the Java API documentation - `PrintStream` class method summary**

void	print(char c) Prints a character.
void	print(char[] s) Prints an array of characters.
void	print(double d) Prints a double-precision floating-point number.
void	print(float f) Prints a floating-point number.
void	print(int i) Prints an integer.
void	print(long l) Prints a long integer.
void	print(Object obj) Prints an object.
void	print(String s) Prints a string.
PrintStream	printf(Locale l, String format, Object... args) A convenience method to write a formatted string to this output stream using the specified format string and arguments.
PrintStream	printf(String format, Object... args) A convenience method to write a formatted string to this output stream using the specified format string and arguments.
void	println() Terminates the current line by writing the line separator string.
void	println(boolean x) Prints a boolean and then terminate the line.
void	println(char x) Prints a character and then terminate the line.
void	println(char[] x) Prints an array of characters and then terminate the line.
void	println(double x) Prints a double and then terminate the line.
void	println(float x) Prints a float and then terminate the line.
void	println(int x) Prints an integer and then terminate the line.
void	println(long x) Prints a long and then terminate the line.
void	println(Object x) Prints an Object and then terminate the line.
void	println(String x) Prints a String and then terminate the line.

Source: https://docs.oracle.com/en/java/javase/23/docs/api/java.base/java/io/PrintStream.html

In the Java API documentation, there are several overloaded methods in most of the Java classes. For example, in Figure 3.7, if you will examine the `PrintStream` class under the `java.io` package, you will notice that there are several declarations of the `print()` and `println()` methods, all having a different argument values and because of this we can conveniently use the `println()` method in printing any value, may it be an `int`, a `double`, or even a `String` value. If we need to print a value, we just call

System.out.println() method and Java will then match the value that was passed to the println() method to the corresponding println() method in the PrintStream class. No need to check on the argument type. Because the print() and the println() methods are overloaded methods that can accommodate any type of data that needs to be printed.

Overloaded methods can be declared within the same class or within the same class hierarchy. This feature enables us to have a convenient way of writing and calling methods that may have a different set of arguments.

3.10 Rules for Method Overloading

If you want to write your own overloaded method, you may want to consider the following rules:

1. All overloaded methods must have the same method name.
2. All overloaded methods must have a different parameter/argument list.
3. As for the return type, all overloaded methods may or may not have the same return type.

In the sample code below, we have created several versions of the sum() method that is supposed to compute the sum of all integer values passed as a parameter. Here's a sample code that shows method overloading:

```
public class Overloading1 {

  public static void main(String[] args){
```

```
4
5      System.out.println(sum(10, 20));
6      System.out.println(sum(10, 20, 30));
7      System.out.println(sum(10, 20, 30, 40));
8    }
9    public static int sum(int num1, int num2) {
10
11     return num1 + num2;
12   }
13   public static int sum(int num1,
14                         int num2, int num3) {
15     return num1 + num2 + num3;
16   }
17   public static int sum(int num1, int num2,
18                         int num3, int num4) {
19
20     return num1 + num2 + num3 + num4;
21   }
22 }
```

In this example, we see that there are several declarations of the sum method, all of which can be seen in lines 9, 13, and 17.

If the user then needs to add up 2, 3, or 4 integer values, all we need to do is to call a `sum()` method and pass the integer values as arguments, all the declared `sum()` methods will accept the `int` values passed to it and compute for the `sum` and return the integer `sum` of all parameters passed to it. The only limitation is that the `sum()` methods in the given code can only accept 2, 3, or 4 `int` values. It would be impractical to write a method for every number of `int` values.

To overcome limitations like this, a new language feature was introduced in Java SE 5.0, it is called *var-args* (variable arguments). This means that any number of arguments can now be passed to a var-args method.

Every new language feature comes with a handful of rules. Here are the rules for declaring var-args methods:

1. **There can only be one var-args per method.**

Correct:

```
protected void doStuff(double... args){}
public static void main(String... args){}
public void doSomething(int... args){}
```

Incorrect:

```
protected void doStuff(double... arg1, int... args2){}
// Java does not allow multiple varargs parameters in
// a single method. Only one varargs parameter is
// allowed, and it must be the last parameter.

public static void main(String... args[]){}
// The syntax incorrectly combines varargs (...)
// and array notation ([])

public void doSomething(int[] args...){}
// Placing the varargs (...) after the array
// is incorrect. The proper way is int... args.
```

2. **If there are multiple parameters, the var-arg parameter must be the last parameter in the list of parameters.**

Correct:

```
protected void doStuff(String str, Object... args){}
public void doSomething(char ch, int... args){}
```

Incorrect:

```
protected void doSomething (double... arg1, int args2){}
// Varargs (...) must be the last parameter in a
// method signature. Here, double... arg1 is
// incorrectly placed before int args2.
public void doStuff(String param1, char... param2, int param3){}
// Similar issue, char... param2 should be the
// final parameter, but an int parameter follows
// it. Varargs can only appear at the end.
```

Here's a sample code that uses var-args:

```
public class Overloading2 {

  public static void main(String[] args) {

      System.out.println(sum(10, 20));
      System.out.println(sum(10, 20, 30));
      System.out.println(sum(10, 20, 30, 40));
      System.out.println(sum(1.0, 2.0, 3.0, 4.0));
  }
  public static int sum(int... num) {

      int total = 0;
      for (int n : num) {

```

```
15          total += n;
16      }
17      return total;
18  }
19
20  public static int sum(double... num) {
21
22      double total = 0;
23      for (double n : num)
24      {
25          total += n;
26      }
27      return total;
28  }
29 }
```

In line 10, the argument list is now using the *var-args* notation. The method `sum()` can now accept any number of `int` values in the method `sum()`. It can now accept 2 numbers, 3 numbers, 4 numbers, 10 numbers, and even a hundred numbers. After the `sum()` method receives all the `int` values, these values will be treated as an array of `int` inside the `sum()` method. There's also line 20, where we can accept any number of double values in the method `sum()`.

A common mistake in calling the var-args method is when an array value is passed. Typically, they accept primitive data types and then implicitly convert them into an array once the values are read inside the var-args method, as you can see in line 12, where an enhanced for loop was used to traverse the values.

3.11 Inheritance and Constructors

We mentioned that inheritance is a feature that allows you to carry over methods and variables to a subclass. This is true only for methods and variables, but not with constructors. In Java, constructors are not inherited by subclasses.

Consider this example:

```
// ConstructorChaining.java
class A {
    A() {
            System.out.println("A");
    }
}
class B extends A {
    B() {
            System.out.println("B");
    }
}
class C extends B {
    C() {
            System.out.println("C");
    }
}
public class ConstructorChaining {
```

```
    public static void main(String[] args) {

        C c = new C();
    }
}
```

After instantiating the object of class `C`, you will notice that the output will be:

```
Output:
A
B
C
```

The output gives the impression that the superclass constructors were inherited; however, this is untrue.

There is a rule with regard to inheritance and constructors. **Constructors are not inherited, but they are chained**. We have what we call constructor-chaining.

The given example above is transformed after compilation by inserting the `super()` keyword in the first line of each constructor.

Here is the updated version of the given code. In this version, Java has already made some changes.

```
class A extends Object {

    A() {

        super();
        System.out.println("A");
    }
```

```
}
class B extends A {

    B() {

        super();
        System.out.println("B");
    }
}
class C extends B {

    C() {

        super();
        System.out.println("C");
    }
}

public class ConstructorChaining {

    public static void main(String[] args) {

        C c = new C();
    }
}
```

Once compiled and run, it will have the output:

```
A
B
C
```

This code will have the same output as the previous one.

Notice lines 5, 13, and 21. The keyword `super()` was inserted in the first line of each constructor.

As a rule, if a constructor does not have a call to `this()` or a call to `super()`, Java will add a default call to `super()` in the first line of the constructor, just like what you had in lines 5, 13, and 21. The call to `super()` allows the current constructor to call the constructor in the superclass. This is where the term constructor-chaining was coined.

How about the changes we had in line 1?

```
1  class A extends Object
```

Another rule. If a class does not extend any superclass, it automatically extends the class `Object`. All classes in the API and the user-defined classes will have this rule. The class `Object` is basically the root of all class hierarchies.

3.12 Overloading Constructors

In Chapter 2, we had a discussion about Constructors. We said that constructors are method-like constructs and, just like methods, constructors can also be overloaded. Here is a sample code that demonstrates constructor overloading.

```
// Employee.java
public class Employee {
  private String name;
  private MyDate birthDate;
  private double  salary;
  private static final double BASE_SALARY = 15000.00;
```

```
 // Constructor
public Employee(String name, double salary,
                              MyDate DoB) {

    this.name = name;
    this.salary = salary;
    this.birthDate = DoB;
}
public Employee(String name, double salary) {

    this(name, salary, null);
}
public Employee(String name, MyDate DoB) {

    this(name, BASE_SALARY, DoB);
}
public String getDetails() {

 return "Name: " + name + "\nSalary: "
     + salary + "\nBirth Date: " + birthDate;
}
}
```

The `Employee` class has three constructors, just like the rule for method overloading, all constructors must have different argument lists.

Using `this()` or `super()` is optional for all constructors. The call to `this()`, is actually a call to a fellow constructor or a constructor within the same class, while a call to `super()` is a call to the constructor of the superclass.

The `Employee` class has three constructors, just like the rule for method overloading, all constructors must have different argument lists.

Using `this()` or `super()` is optional for all constructors. The call to `this()`, is actually a call to a fellow constructor or a constructor within the same class, while a call to `super()` is a call to the constructor of the superclass.

If the call to `this()` or the call to `super()` is used, they must be placed on the first line of the constructor. Placing them elsewhere will cause a compilation error. But in the absence of either of the two, a default call to `super()` will be placed on the first line of the constructor.

Here is the `Manager` class, which demonstrates the call to `super()`.

```
// Manager.java
public class Manager extends Employee {

    private String dept;

    public Manager(String name,
                double salary, String dept) {
        super(name, salary);
        this.dept = dept;
    }
    /* This code will fail. A default call
       to super() will be in place
    public Manager(String dept) {
       this.dept = dept;
    }
    */
    // additional Manager code...
}
```

In line 7 of the `Manager` class, a call to `super` passes the `name` and `salary` values to the constructor in the `Employee` class. This will allow you to reuse one of the constructors in the `Employee` class, thus removing the possibility of having redundancies in your code.

As for line 12 of the `Manager` class, it will cause an error because the first line of the constructor does not have a call to `this()` or a call to `super()`, therefore Java will put a call to `super()` in the first line of the constructor, telling it to look for a constructor in its superclass (in this case, the class `Employee`) to run a constructor that does not have any arguments.

3.13 The `Object` Class

The `Object` class is the root of all classes. Whether it is a pre-written class in the API or a user-defined class, ALL classes are a subclass of the `Object` class.

And because of these, all methods in the `Object` class are implicitly inherited by all classes, so there is no need to use "`extends Object`". Methods like `equals()`, `hashCode()`, and the `toString()` method.

Consider the class `Person`.

```
// Person.java
public class Person extends Object {
 private String name;
 private int age;
 public Person(String name, int age) {
       this.name = name;
```

```
        this.age = age;
    }
    // getters
    public String getName() {
        return name;
    }
    public int getAge() {
        return age;
    }
}
```

Since the class `Person` is a user-defined class, there is still no user-defined implementation of the `equals()`, `hashCode()`, and `toString()` methods. The current implementation of these methods is the one inherited from the `Object` class.

In Java, comparing objects using the `==` operator and the `.equals()` method can produce different results. By default, they have the same behavior. The `equals()` method is inherited from the `Object` class and compares based on the == operation. The `==` operator checks if two object references point to the same memory location, while the `.equals()` method is typically used to compare the content or logical equality of objects. The `equals()` method is typically overridden when creating a user-defined class.

In Java, comparing objects using the `==` operator and the `.equals()` method can produce different results. The `==` operator checks if two object references point to the same memory location, while the `.equals()` method is typically used to compare the content or logical equality of objects.

The following code demonstrates this concept by creating two Person objects with identical attributes and comparing them using both approaches.

Consider the given code snippet below that creates two `Person` objects, `p1` and `p2`, and compares them using the == operator and the `equals()` method.

```
7   Person p1 = new Person("Lawrence", 30);
8   Person p2 = new Person("Lawrence", 30);
9
10  System.out.println(p1 == p2);
11  System.out.println(p1.equals(p2));
```

The output of lines 10 and 11 will be:

```
false
false
```

Ideally, since `p1` and `p2` have the same name and age, the output should be `true`. But since the implementation of the `equals()` method came from the `Object` class, we will have both statements print `false` values because the `equals()` implementation of the `Object` class is the same as the == implementation.

As for the `hashCode()` method output of both objects, it can be printed by inserting this:

```
14  System.out.println(p1.hashCode());
15  System.out.println(p2.hashCode());
```

A possible output of lines 14 and 15 will be:

```
2018699554
1311053135
```

The hashCode() method is tied to object identity by default, unless you explicitly override it. If you do not override the hashCode() in your class, then two different objects (p1 and p2), even if they have the same attributes and values, will usually produce different hash codes because the default hashCode() is based on the object's memory addresses, as seen in the output above.

If you do override hashCode() properly (along with equals()), then objects with the same attribute values should return the same hash code. This is important for collections like HashMap, HashSet, and Hashtable to work correctly.

If two objects are equal according to equals(), they must return the same hashCode(). But if two objects have the same data without overriding equals() and hashCode(), they are still considered different objects with potentially different hash codes.

Another problem with the current Person objects is the String representations of both p1 and p2. If we try to print the values of p1 and p2 by doing this:

```
System.out.println(p1);        // same as //p1.toString()
System.out.println(p2.toString());
```

We will have this output:

```
Person@7852e922
Person@4e25154f
```

These are actually the hexadecimal values of their generated hashCode values.

So far, the method implementation we got from the inherited equals(), hashCode(), and toString()

methods from the `Object` class is not the one that we needed.

To provide our own implementation of these methods, we need to override the `equals()`, `hashCode()`, and `toString()` methods from the `Object` class. Consider the updated version of the `Person` class.

```
// Person.java
public class Person extends Object {

    private String name;
    private int age;

    // constructor
    public Person(String name, int age) {

        this.name = name;
        this.age = age;
    }

    // getters
    public String getName() {

        return name;
    }
    public int getAge() {

        return age;
    }

    // override the equals method
    @Override
```

```
    public boolean equals(Object obj) {

        boolean result = false;
        if (obj != null && obj instanceof Person) {

            Person p = (Person) obj;
            if (this.getName().equals(p.getName())
             && this.getAge() == p.getAge()) {

                result = true;
            }
        }
        return result;
    }

    // override the hashCode method
    @Override
    public int hashCode() {

        return getName().hashCode() ^ getAge();
    }

    // override the toString method
    @Override
    public String toString() {

        return getName() + ", at age " + getAge();
    }
}
```

To test the output, we create a test class called TestPerson.java with the following code:

```
// TestPerson.java
public class TestPerson {

      public static void main(String[] args) {

            Person p1 = new Person("Lawrence", 30);
            Person p2 = new Person("Lawrence", 30);

            System.out.println(p1 == p2);
            System.out.println(p1.equals(p2));

            System.out.println(p1.hashCode());
            System.out.println(p2.hashCode());

            System.out.println(p1);
            System.out.println(p2.toString());
      }
}
```

If we compile and run this code, we should have this output:

```
false
true
2314851
2314851
Lawrence, at age 30
Lawrence, at age 30
```

The first two lines in the output:

```
false
true
```

With the `equals()` method implementation in the `Person.java`, these are outputs for lines 9 and 10 of the `TestPerson.java` code.

```
9       System.out.println(p1 == p2);
10      System.out.println(p1.equals(p2));
```

The `hashCode()` values of both `p1` and `p2` are now identical because of the `hashCode()` implementation in the `Person` class.

As for the `toString()` method, you can override the `toString()` method of the `Object` class if you want your object to be printed using your preferred format.

Chapter Summary

- Introduction to Inheritance was discussed, which was one of the fundamental pillars of OOP. This promotes code reuse.
- Classes, superclasses, and subclasses are essential concepts that one should know when talking about the concepts of inheritance. In other OOPs, certain terms can be used instead, like parent and child classes. But in Java, we use the terms superclass and subclass.
- The "is-a" relationship is another term that is also used in software engineering. In some cases, we use the term *'a subclass of'* or *'derived from'*.
- Java is unique among other object-oriented programming languages in that it uses the concept of single inheritance. However, we will talk about how to use the ideas of multiple inheritance in Java in Chapter 5: Other Class Features.
- In chapter two, we have introduced the `public` and `private` access modifiers to explain the concepts of encapsulation, but in this chapter, we have discussed all four access modifiers.
- Method overriding in Java occurs when a subclass provides a specific implementation of a method that is already defined in its superclass. This allows the subclass to customize or completely replace the behavior of that method while maintaining the same method signature. There are five method overriding rules discussed in this chapter.
- The `super` keyword is used within a subclass to refer to its immediate superclass. It can be used to call the superclass's constructor, access superclass methods, or access superclass fields, thereby allowing the subclass to inherit and utilize the functionality of its superclass.

- Overloading methods, sometimes called method overloading in other resources, allows a class to have multiple methods with the same name but different parameters. This provides flexibility by enabling different ways to call a method based on the number or types of parameters passed, improving code readability and reusability within the same class. We also discussed the rules for method overloading in this chapter.
- In Java, inheritance and constructors are closely related but have distinct roles. When a subclass is instantiated, its constructor implicitly or explicitly calls the constructor of its superclass to ensure proper initialization of the inherited fields. This is typically done using the super keyword, which must be the first statement in the subclass constructor if used to call the superclass constructor. We also looked at the use of the `super()` and the use of `this()` in this chapter.
- Overloading constructors in Java allows a class to have multiple constructors with different parameter lists. This provides flexibility in creating objects with various initial states, as each constructor can perform different initialization tasks based on the arguments passed during object creation.
- The `Object` class, which is the root of all classes, was also discussed. It is also suggested that when you are creating your own class, it is better to override the `toString()`, `equals()`, and `the hashCode()` methods in your own user-defined class.

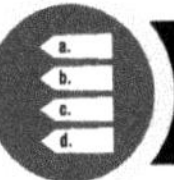

Multiple Choice Questions

1. **Which of the following is NOT TRUE about constructors?**
   a. We can overload constructors.
   b. We can reuse a constructor.
   c. We can chain constructors.
   d. We can inherit constructors from a superclass.

2. **Consider the following code:**

```
class MyClass  {
    MyClass(String str1, String str2)  {
          System.out.print("Java  ");
    }
    MyClass(String s1, String s2)  {
         System.out.print("Oracle  ");
    }
    public static void main(String args[]) {
         new MyClass("Java", "Oracle");
    }
}
```

What will happen if you compile the code?

a. The code will compile and run and will display: Java Oracle.
b. The code will cause a compilation error because the constructors have identical parameters.
c. The code will compile and run and will display: Java.
d. The code will compile and run and will display: Oracle.

3. Given the class below:

```
class A {
    public void printInfo() {
        System.out.print("A");
    }
}
class B extends A {
    public void printInfo() {
        System.out.print("B");
    }
}
class Test {
    public static void main(String args[]) {
        A a = new A();
        B b = new B();
        a.printInfo();
        b.printInfo();
    }
}
```

What is the output when we compile and run the code?

a. AB
b. A
c. B
d. Compilation Error

4. **The ________ class, defined in the java.lang package defines and implements behavior common to all classes, including the user-defined classes.**
   a. `Object`
   b. `Objects`
   c. `String`
   d. `Integer`

5. **Which of the following is NOT a method that can be overridden when creating a new class?**
   a. extends()
   b. toString()
   c. equals()
   d. hashCode()

6. **Which of the following statements is NOT TRUE?**
   a. If a class will inherit another class, you need to use the keyword inherit to declare the name of the superclass.
   b. A class can only have a single superclass.
   c. All classes are a subclass of the `java.lang.Object` class.
   d. A class may have multiple subclasses.

7. **Which of the following statements is NOT TRUE about method overriding?**
   a. The return type of the overriding method should be the same or a subtype of the return type declared in the original overridden method in the superclass.
   b. Methods that you override should have the same spelling, and it is NOT case-sensitive
   c. The argument list should be exactly the same as that of the overridden method.
   d. The access level cannot be more restrictive than the overridden method's access level.

8. **Given two classes, the** `Person` **class and the** `Employee` **class:**

```
class Person  {
    private String name;
}
class Employee extends Person  {
    private double salary;
    public static void main(String[] args) {
    Employee e = new Employee();      // 1
    e.name = "Peter Parker";          // 2
    e.salary = 1234.56;           // 3
    System.out.println("Peter Parker has a salary of: "
                                + e.salary); //4
    }
}
```

Which line will cause a compilation error?

a. Line 1 will cause a compilation error.
b. Line 2 will cause a compilation error.
c. Line 3 will cause a compilation error.
d. Line 4 will cause a compilation error.

9. **Which of the following statements is NOT TRUE about Constructor chaining?**
   a. In the absence of the call to `this()` and call to `super()`, Java will automatically insert the call to `super()` in the first line of the constructor.
   b. The call to `super()` is used to call the constructor of the superclass.
   c. Constructors CANNOT be overloaded.
   d. The call to `this()` is used to call the constructor of the current class.

10. **Which of the access modifiers can limit the accessibility within the same class, same package, and same inheritance tree?**
    a. `private`
    b. <default>
    c. `protected`
    d. `public`

Multiple Choice Answer Key

1 – d	2 – b	3 – a	4 – a	5 – a
6 – a	7 – b	8 – b	9 – c	10 – c

Coding Tasks

Coding Task 1:

Given the UML Class diagram with the list of attributes and a constructor. Write the class version of the UML diagram and save it as `Student.java`. Together with the Student.java file, you also have to create the corresponding test class named: `TestStudent.java`, which contains the `main` method.

```
Student
-----------------------------
- name: String
- studentId: int
- course: String
-----------------------------
+ Student (String name,
           int studentId,
           String course)
//getters()
//toString()
//equals()
//hashCode()
```

In the main method, you are to create two `Student` objects named `s1` and `s2`. Provide an identical `name`, `studentId`, and `course` to both objects for the purpose of testing the `equals()` and `hashCode()` methods. Override the `toString()` method as well to print out the identity of the student objects. Provide the setter and the getter methods as well.

Coding Task 2: Implement a Simple Inheritance Hierarchy

Objective:

Understand and apply inheritance by creating a basic hierarchy of classes with common and specific behaviors.

Instructions:

A. Create a base class `Animal`:

Fields:

`name (String)`: The name of the animal.

`age (int)`: The age of the animal.

Constructor:

Initialize name and age.

Methods:

`makeSound()`: A method that prints a generic sound ("Some generic animal sound").

`displayInfo()`: A method that prints the animal's `name` and `age`.

B. Create a subclass `Dog` that extends `Animal`:

Constructor:

Call the superclass constructor to initialize the `name` and `age`.

Methods:

Override the `makeSound()` method to print "`Woof! Woof! Woof!`".

Add a method `fetch()` that prints "`Fetching the ball!`".

C. Create another subclass, `Cat`, that extends `Animal`:

Constructor:

Call the superclass constructor to initialize the `name` and `age`.

Methods:

Override the `makeSound()` method to print "`Meow! Meow! Meow!`".

Add a method `purr()` that prints "`Purring softly!`".

D. Create a `TestAnimal` class with the `main` method:

Instantiate objects of `Dog` and `Cat`.

Call their methods to demonstrate the behavior of each class.

Here's the code for your TestAnimal class:

```
public class TestAnimal {
    public static void main(String[] args) {
        Dog myDog = new Dog("Buddy", 3);
        Cat myCat = new Cat("Whiskers", 2);

        myDog.displayInfo();
        myDog.makeSound();
        myDog.fetch();

        myCat.displayInfo();
        myCat.makeSound();
        myCat.purr();
    }
}
```

Here's the expected output:

```
Name: Buddy, Age: 3
Woof! Woof! Woof!
Fetching the ball!
Name: Whiskers, Age: 2
Meow! Meow! Meow!
Purring softly!
```

CHAPTER 4
Polymorphism

Key Learning Objectives

- Learn the concept of polymorphism
- Virtual method invocation
- Heterogeneous array
- Methods with polymorphic arguments
- The `instanceof` operator
- Casting of objects

This chapter introduces the concept of polymorphism, one of the core features in object-oriented programming (OOP) that allows objects of different classes to be treated as instances of a common superclass. We'll explore how virtual method invocation enables dynamic method resolution at runtime, enhancing flexibility in code. You'll learn how to work with heterogeneous arrays and use methods that accept polymorphic arguments, broadening the scope of what functions can handle. Additionally, we will cover the `instanceof` operator for type-checking and how object casting is used to convert between different types within an inheritance hierarchy.

4.1 Introduction to the Concept of Polymorphism

Polymorphism was derived from two Greek words, "poly", which means many, and "morph", which means forms. So, polymorphism can literally mean "having many forms". In Java, the object itself is non-polymorphic, while a reference variable that holds the object is the one that is polymorphic. This means that in Java, an object is just a specific thing created from a class, like a car made from a blueprint. That object is not flexible—it is what it is.

However, a reference variable (the variable that points to the object) can be flexible. If the reference variable is of a parent class, it can point to different types of objects that belong to its family (like a Vehicle variable pointing to a Car or Bike). This is called polymorphism—the ability of one reference type to represent different object types. Think of it like a universal remote control that can control different devices. The remote is like the reference variable, and the devices (TV, AC, DVD player) are like the objects.

In our example for this chapter, I'll be using the `Employee` class, referring to different types of Employees like Managers, Directors, Engineers, and Secretaries. Let us consider the hierarchy in Figure 4.1 again.

Figure 4.1 Example of single inheritance hierarchy

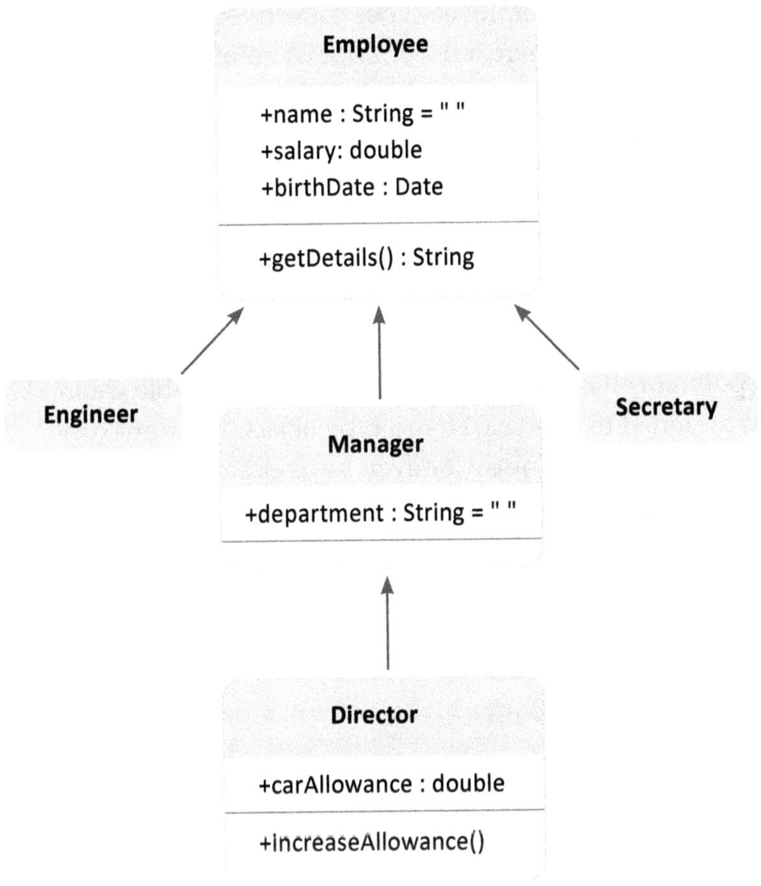

Here's an example of how polymorphism is used.

```
Employee e = new Manager();           // this is allowed
e.department = "Sales";               // this is illegal
System.out.println(e.getDetails());   // this is ok!
```

Storing an instance of the `Manager` class as an `Employee` type leverages polymorphism, allowing you to treat all employees uniformly regardless of their specific roles. Since `Manager` is a subclass of `Employee`, it inherits the basic characteristics of an employee while adding unique features. This is particularly useful in scenarios where you need to manage various employee types collectively, such as in a payroll system where a list of `Employee` objects may include managers, developers, and interns. By generalizing behavior through the `Employee` type, your code becomes more flexible and easier to maintain. If a specific behavior unique to the `Manager` is needed, the object can be downcast to its original type. This approach simplifies code while enabling dynamic behavior based on the actual object type.

Line 5 is allowed because the reference variable `e` is a polymorphic variable, which means the variable `e` can be assigned to any `Employee` type object, like `Manager`, `Director`, `Engineer`, or even `Secretary`.

Consider the class hierarchy in Figure 4.1. The following Java statements can be substituted in line 5:

```
Employee e = new Employee();
Employee m = new Manager();
Employee d = new Director();
Employee s = new Secretary();
Employee en = new Engineer();
```

These are all legal declarations because all of these classes are subclasses of the `Employee` class. As long as there is a class that is a subclass of `Employee`, you can assign it to the reference variable `Employee e`.

To easily recall this rule, you can follow this syntax:

```
<superclass> var = new <class or subclass>;
```

However, in line 6, this statement will cause a compilation error.

```
6    e.department = "Sales";                // this is illegal
```

This is because upon compilation, the `department` attribute is not found in the class `Employee`. The `department` attribute can only be found in the class `Manager` and, by virtue of inheritance, also in its subclasses.

When the `e.department` call happened in line 6, the `department` attribute that was only declared in the `Manager` class was checked by the compiler, and it was not found in the `Employee` class.

During compilation, the left-hand side of an assignment (=) operator was checked and evaluated, while the right-hand side was evaluated at runtime.

On the left-hand side of the assignment operator (=), we have;

```
5    Employee e = new Manager();
```

The statement **`Employee e`** was evaluated at compile time. At the same time, the next statement in line 6 was also evaluated;

```
6    e.department = "Sales";
```

It will cause a compilation error because the reference variable `e` can only access the `name`, `salary`, `birthDate` attributes, as well as the `getDetails()` method, which are members of the `Employee` class. The `department` attribute is not a member of the `Employee` class, and therefore caused the compilation error.

4.2 Virtual Method Invocation: The `getDetails()` Method

Consider the code snippet:

```
5   Employee e = new Manager();
        ....
7   System.out.println(e.getDetails());
```

In line 7, the `getDetails()` method was called, which is present in both the `Employee` and `Manager` classes. The `Employee` class has a declared `getDetails()` method, while the `Manager` class inherited it (and possibly overrides it).

At runtime, the `getDetails()` method that will execute is the one that is associated with the object. In this example, the `getDetails()` method that will run is the version declared under the `Manager` class. This process is called the Virtual Method Invocation (VMI).

The `Employee` class's implementation of the `getDetails()` method will only return the `String` representation of the `name` and `salary` of the object, while the `Manager` class's `getDetails()` method implementation will return something more. It can return a `String` representation of the `name` and `salary`, and the `department` values of the `Manager` object.

This polymorphic behavior allows the method implementation to change at runtime. The method name must be a member of the declared class—in this case, the `Employee` class is checked at compile time. While at runtime, the method implementation of the `getDetails()` method will come from the instantiated object, which, in this case, comes from the `Manager` class.

Here's a working code that contains additional methods that will demonstrate polymorphism better. The source code, `Employee.java` that contains all three classes: the `Employee` class, the `Manager` class, and the `Director` class.

```
class Employee {
  String name;
  double salary;
  public String getDetails() {
     return "Name: " + name +
        "\nSalary: " + salary;
  }
  public void work() {
     System.out.println("Work work work...");
  }
}
class Manager extends Employee  {
  String department;
  public String getDetails() {
     return super.getDetails() +
           "\nDepartment: " + department;
  }
  public void callMeeting() {
     System.out.println("Call Department Meeting..");
  }
}
class Director extends Manager  {
  double carAllowance;
  public String getDetails() {
```

```
        return super.getDetails() +
                "\nCar Allowance: " + carAllowance;
    }
    public void playGolf() {
        System.out.println("Play Golf..");
    }
}
```

Together with the Employee.java, we also have the TestEmployee.java. Here's the code:

```
public class TestEmployee  {
    public static void main(String[] args)    {
        Employee e = new Employee();
        Employee m = new Manager();
        Employee d = new Director();

        doEmployeeStuff(e);        // an Employee object
        // doEmployeeStuff(m);     // an Manager object
        // doEmployeeStuff(d);     // an Director object
    }

    public static void doEmployeeStuff(Employee e)
    {
        e.work();
        if (e instanceof Director)
        {
                Director d = (Director) e;
                d.callMeeting();
                d.playGolf();
        }
```

```
21          else if (e instanceof Manager)
22          {
23                  Manager m = (Manager) e;
24                  m.callMeeting();
25          }
26      }
27  }
```

If the doEmployeeStuff() method takes in an Employee e object from line 7, the output will be:

```
Work work work...
```

If changed to a Manager object by using line 8's doEmployeeStuff(m); then the output will be:

```
Work work work...
Call Department Meeting..
```

And if changed to a Director object by using line 9's doEmployeeStuff(d); then the output will be:

```
Work work work...
Call Department Meeting..
Play Golf..
```

4.3 Heterogeneous Array

In most programming languages, arrays are usually homogeneous in nature. A homogeneous array will always have element values of the same type. Consider this array of `String` objects that represents our favorite coffee concoction. This is an example of a homogeneous array.

```
String coffee[] = {"Espresso", "Java", "Mochachino"};
```

With polymorphism, a heterogeneous array is also possible. Consider the hierarchy in Figure 4.1 given earlier. Let us say you have an array of `Employee` objects called `staff[]`, and your boss wants you to display all the details of all the employee objects, which include a variety of different employee objects like `Director`, `Manager`, `Secretary`, `Employee`, and `Engineer` objects.

Consider this code snippet:

```
// declares an array of Employees called staff
// that demonstrates a heterogeneous array
Employee staff[] = new Employee[5];
// each element of the array staff[] may be
// assigned to a different type of Employee object.
// With polymorphism, this is possible as long as
// the object type is a subclass of the Employee class
staff[0] = new Manager();
staff[1] = new Director ();
staff[2] = new Employee();
staff[3] = new Secretary();
staff[4] = new Engineer();
for (int i = 0; i < staff.length; i++)
{
   System.out.println(staff[i].getDetails());
}
```

This sample code will display all the details of all your employees by calling the `getDetails()` method of each object.

Every time the loop iterates, the variable `i` increments starting from 0, and the statement **`staff[i].getDetails()`** will return different values depending on what object **`staff[i]`** is referring to.

If `i = 0`, then `staff[0]` is referring to a `Manager` object, and it will return the `name`, the `salary`, and the `department` information for the `Manager` object.

If `i = 1`, then `staff[1]` is referring to a `Director` object and will return the `name`, the `salary`, the `department`, and the `carAllowance` information for the `Director` object.

If `i = 2`, then `staff[2]` is referring to an `Employee` object; thus, it will return the `name` and the `salary` information for the `Employee` object.

This is one area where virtual method invocation in Java is useful.

4.4 Methods With Polymorphic Arguments

Another way to demonstrate polymorphism in Java is by declaring methods with *polymorphic arguments*. This feature allows you to create a method that can accept an object of any type within the subclass of the data type in the argument.

Consider the UML Diagram in Figure 4.2 that shows the hierarchy of the `Animal` class.

Figure 4.2 Single Inheritance Hierarchy Example for Animal Class Hierarchy

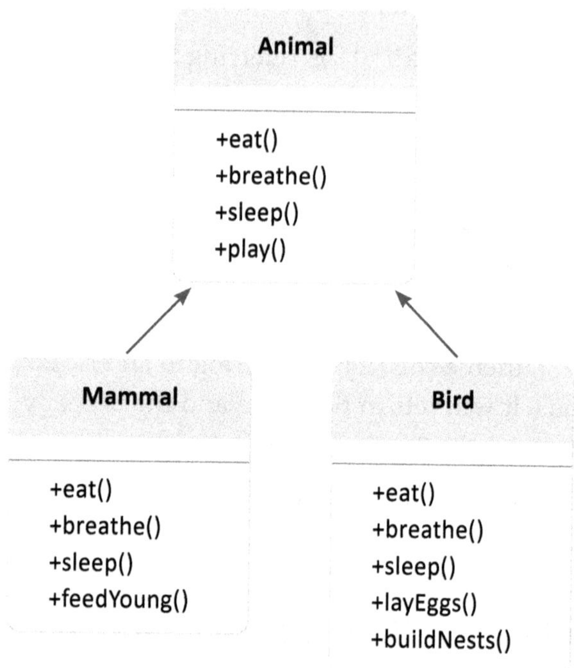

The `Animal` class has two subclasses, the `Mammal` and the `Bird` class. Each subclass has its own unique methods. For the `Mammal` class, it has a unique method of `feedYoung()`, while the `Bird` class has a unique set of methods, `layEggs()` and `buildNests()`.

Here are the sample codes for `Animal`, `Mammal`, and `Bird` classes.

```
// Animal.java
public class Animal {
  // some animal-based methods
  public void eat() { some implementation here }
  public void breathe() { some implementation here }
```

```
  public void sleep() { some implementation here }
  public void play() {
        System.out.println("Have fun.");
  }
}
```

```
// Mammal.java
public class Mammal extends Animal {
  // override all animal-based methods
  @Override
  public void eat() {
        System.out.println("Mammal Eats");
  }
  @Override
  public void breathe() {
        System.out.println("Mammal Breathe");
  }
  @Override
  public void sleep() {
        System.out.println("Mammal Sleeps");
  }

  // my own mammalian methods
  public void feedYoung() {
        System.out.println("Feed Young");
  }
}
```

```
// Bird.java
public class Bird extends Animal {

  // override all animal-based methods
```

```
    @Override
    public void eat() {
          System.out.println("Bird Eats");
    }
    @Override
    public void breathe() {
          System.out.println("Bird Breathe");
    }
    @Override
    public void sleep() {
          System.out.println("Bird Sleeps");
  }

    // my own birdly methods
    public void layEggs() {
          System.out.println("Lay Eggs");
    }
    public void buildNest() {
          System.out.println("Build Nests");
    }
}
```

Here is a sample code that demonstrates the use of polymorphic arguments for methods:

```
// TestAnimal.java
public class TestAnimal {

    public static void main(String[] args) {

          Mammal dog = new Mammal();
          Bird eagle = new Bird();
```

```
8
9           doAnimalStuff(dog);
10          doAnimalStuff(eagle);
11      }
12      public static void doAnimalStuff(Animal a) {
13
14          a.eat();
15          a.breathe();
16          a.sleep();
17          a.play();
18
19          if (a instanceof Mammal) {
20
21                  Mammal m = (Mammal)a;
22                  m.feedYoung();
23          }
24          else if (a instanceof Bird) {
25
26                  Bird b = (Bird) a;
27                  b.layEggs();
28                  b.buildNest();
29          }
30      }
31  }
```

In lines 9 and 10, the method `doAnimalStuff()` passes a `dog` and an `eagle` object.

```
9   doAnimalStuff(dog);
10  doAnimalStuff(eagle);
```

Line 12 of the code `TestAnimal.java`, the method `doAnimalStuff(Animal a)` accepts any type of `Animal` object as its parameter. You may pass a `Mammal` object or a `Bird` object in the said method. This is possible as long as the objects that you pass are subclasses of the class `Animal`. In this case, the `Mammal` and `Bird` classes are subclasses of the `Animal` class. This is an example of how polymorphic arguments are used.

4.5 The `instanceof` Operator

The `instanceof` operator can be used to test what type of object is assigned to a reference variable.

Given lines 19 and 24, the `instanceof` operator was used with the `if-else` or the `if-else-if` ladder to test if variable `a` is assigned to an object of a specific data type.

In this case, line 19 is testing if the object assigned to variable `a` is an object instance of class `Mammal`. While on line 24, the variable `a` is being tested to check if it is assigned to an object instantiated from the class `Bird`.

Using the `instanceof` operator, if the passed argument is an "instance of" a certain class that is tested, it will return either a `true` or a `false` value.

```
19      if (a instanceof Mammal) {
20
        . . .
23      }
24      else if (a instanceof Bird) {
25
        . . .
29      }
```

The `instanceof` operator is a type of relational operator that yields a boolean value.

4.5.1 Pattern Matching With `instanceof` (JDK 14 and Later)

The `instanceof` operator in Java is used to check whether an object belongs to a particular type. Before JDK 14, whenever you used `instanceof`, you usually had to follow it with an explicit cast before working with the object. This made code longer and slightly harder to read.

In the old way (Before Java 14):

```
if (obj instanceof String) {
    String text = (String) obj; // manual casting
    System.out.println(text.toLowerCase());
}
```

Starting in JDK 14 (and finalized in JDK 16), Java introduced pattern matching for `instanceof`. This feature lets you check the type and assign the result to a variable in a single step. The variable is automatically cast, so you can use it right away without extra code.

New Way (JDK 14+):

```
if (obj instanceof String text) {
    System.out.println(text.toLowerCase());
}
```

Here, `text` is called a *pattern variable*. If the condition is `true`, Java automatically treats it as the correct type inside the `if` block.

Why is this better?

- Less code: No need for separate casting.

- Clearer intent: The type check and variable creation happen in one place.
- Fewer mistakes: Reduces the chance of casting errors at runtime.

Here's an example for matching a `String`:

```
Object data = "Pattern Matching in Java";
if (data instanceof String str) {
    System.out.println("Length: " + str.length());
}
```

If `data` is a `String`, `str` is created and ready to use without any casting.

Here's another example for matching a custom class:

```
if (shape instanceof Circle c) {
    System.out.println("Radius: " + c.getRadius());
}
```

If `shape` is a `Circle`, `c` gives direct access to its methods and fields.

Pattern matching for instanceof makes type-checking in Java more concise, readable, and safe, especially when working with complex object hierarchies.

We can also combine *Pattern Matching* and *Logical Operators* together:

```
Object obj = "Hello Java";

if (obj instanceof String s && s.length() > 5) {
    System.out.println("String 's' is longer than"
            + " 5 characters: " + s);
}
```

The type pattern in this case is `obj instanceof String s`. It binds `obj` to a new variable `s` of type `String` after determining whether `obj` is an instance of `String`. This type check is then combined with a condition on the newly bound variable `s` (`s.length() > 5`) using the `&&` (short-circuit AND) operator. This makes it possible to carry out type checking and other actions on the object in a more condensed and readable manner.

4.6 Casting of Objects

Lines 14 to 17 use virtual method invocation because the implementation of these methods will depend on what type of object the reference variable `a` currently has.

While other methods that are exclusive to the `Mammal` and `Bird` classes cannot be accessed if we use the `Animal a` reference type, to be able to access the `feedYoung()` method of the `Mammal` class and the `buildNest()` and `layEggs()` method of the `Bird` class, we need to perform an explicit object casting.

The `Animal a` reference variable, can be cast to either a `Mammal` or a `Bird` object, or to any of the `Animal`'s subclass. But for good programming practice, before casting, it is always best to test the reference data type first by using the `instanceof` operator so that a `ClassCastException` can be avoided during runtime. Refer to lines 21 and 26 for object casting.

```
14    a.eat();
15    a.breathe();
16    a.sleep();
17    a.play();
18
```

```
if (a instanceof Mammal m) {
    m.feedYoung();
}
else if (a instanceof Bird) {
    b.layEggs();
    b.buildNest();
}
```

Testing first before applying a data type casting to objects should be done to a reference variable's subclass to prevent a runtime problem called `ClassCastException`. This is why checking the type of object using an `if-else` statement and the `instanceof` operator is considered a good coding practice.

Chapter Summary

- Polymorphism in Java is another core principle of object-oriented programming that allows objects of different classes to be treated as objects of a common superclass, enabling a single interface to represent different underlying forms (data types). This is achieved through method overriding and method overloading, allowing the same method to perform different functions based on the object that invokes it.
- Virtual method invocation in Java occurs when a method call is resolved at runtime rather than compile-time, allowing the Java runtime to determine the appropriate method implementation to invoke based on the actual object's class. This mechanism supports polymorphism by ensuring that overridden methods in a subclass are called even when referenced through a superclass variable.
- A heterogeneous array in Java is an array that can hold objects of different types, all derived from a common superclass. Using polymorphism, each element in the array can be treated as an instance of the superclass, but at runtime, the specific method implementations of the actual object types are invoked, allowing for diverse object behaviors within the same array.
- Methods with polymorphic arguments in Java can accept parameters of a superclass type, allowing for any subclass object to be passed to the method. This enables the method to operate on a variety of objects, with the specific implementation details handled at runtime, providing flexibility and code reusability.
- The `instanceof` operator in Java is used to check whether an object is an instance of a specific class or implements a particular interface. It allows for safe type

casting by verifying the object's type before performing operations specific to that type.

- Casting of objects in Java allows you to convert an object reference of a superclass type to a subclass type, enabling access to subclass-specific methods and fields. This type of casting requires explicit syntax and must be verified with the `instanceof` operator to ensure the object is actually an instance of the target subclass to avoid encountering a `ClassCastException`.

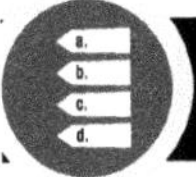

Multiple Choice Questions

1. Given the class declarations:

```
class Person {
    String name;
}
class Employee extends Person {
    double salary;
}
```

Which of the following statements, called independently, will cause a compilation error?

a. Person p = new Employee();
b. Employee e = new Person();
c. Employee e = new Employee();
d. Person p = new Person();

2. Which class relationship should we look for to determine if we can pass an object to a method's parameter?

a. 'just-a'
b. 'is-a'
c. 'has-a'
d. 'but-a'

3. **Consider the class declarations:**

```
class Person
{}
class Employee extends Person
{}
class Manager extends Employee
{}
class Director extends Manager
{}
```

Which of the following declarations is INVALID?

a. Employee d = new Director();
b. Manager m = new Person();
c. Person p = new Manager();
d. Manager md = new Director();

4. **Which of the following statements represents a valid format to call a superclass's constructor from a subclass's constructor?**

a. subclass(name, score);
b. extends(name, score);
c. super(name, score);
d. superclass(name, score);

5. **If jrStudent is an object of Student, which of the following statements will result in a value of true?**

a. instanceof Student = jrStudent
b. Student = instanceof jrStudent
c. jrStudent instanceof Student
d. Student instanceof jrStudent

6. **Given the code below, the line numbers are provided for reference.**

```
1. class SuperClass {
2.   SuperClass doStuff(int x) {
3.     return new SuperClass();
4.   }
5. }
6.
7. class SubClass extends SuperClass {
8. // insert code here
9. }
```

And four declarations:

```
SubClass doStuff(int x) { return new SubClass(); }
SubClass doStuff(int x) { return new SuperClass(); }
SuperClass doStuff(int x) { return new SubClass(); }
SuperClass doStuff(int y) { return new SuperClass(); }
```

How many of the declarations given, inserted independently at line 8, will allow the code to compile?

a. 1
b. 2
c. 3
d. 4

7. **Which of the following is TRUE about Polymorphism?**
    a. Polymorphism is the process of hiding the implementation details from the user by declaring the variables with private accessibility and creating public methods to enable the setting and the getting of the values of the private variables.
    b. Polymorphism is a mechanism of wrapping the data (variables) and code acting on the data (methods) together as a single unit.
    c. Polymorphism lets us reuse the attributes and methods from the parent class.
    d. Polymorphism uses methods either from the parent class or in the current class to perform different tasks. This allows us to perform a single action in different ways.

8. **Consider the class `Device` that contains the method `togglePower()`. The class `Device` has a subclass `Tablet`, which has `browse()`. The class `SmartPhone` has a method `makeCall()` and is a subclass of class `Device`. Which of the following statements IS NOT TRUE?**
    a. SmartPhone "is-a" Tablet
    b. SmartPhone "is-a" Device
    c. Tablet "is-a" Device
    d. Tablet "has-a" togglePower()

9. Consider the class Employee, which has a subclass, the Manager, and the class Director is then a subclass of Manager. Which of the following object instantiations is NOT ALLOWED?

a. A Director d reference type with a Manager object instance.

b. An Employee e reference type with a Manager object instance.

c. An Employee e reference type with a Director object instance.

d. A Manager m reference type with a Director object instance.

10. Given:

```
class A {
 public void printMe() {
    System.out.println("Class A");
 }
}
class B extends A {
 public void printMe() {
    System.out.println("Class B");
 }
}
class C extends B {
 public void printMe() {
    System.out.println("Class C");
 }
}
public class TestClass {
```

```
    public static void doSomething(A a) {
        a.printMe();
    }
    public static void main(String[] args) {
        B b = new B();
        C c = new C();
        doSomething(b);
        doSomething(c);
    }
}
```

What is the output?

a. Class B
Class C

b. Class A
Class A

c. Class A
Class B

d. Class C
Class C

Multiple Choice Answer Key

1 – b	2 – b	3 – b	4 – c	5 – c
6 – c	7 – d	8 – a	9 – a	10 – a

Coding Tasks

Coding Task 1: Student Scholarship

You are hired by your University to create an application that will manage the student records and their scholarship and tuition fee information.

Part of your first project is to set up the hierarchy of the `Student` classes. You have to write 2 Java classes. `Student`, which represents a regular paying student, and `PartialScholar`, which is a subclass of the `Student` class that has a tuition fee discount. The percentage of the scholarship may vary and is sponsored by a certain scholarship provider.

Given the UML, create the respective classes, attributes, constructors, and methods.

For the `Student` class, the constructor should accept the `name` of the student, the `course`, and the `tuitionFee` amount. Create a `static counter` variable and a non-static variable called `studentNumber` that will be used to compute the student number of each created student object and return it in the `toString()` method.

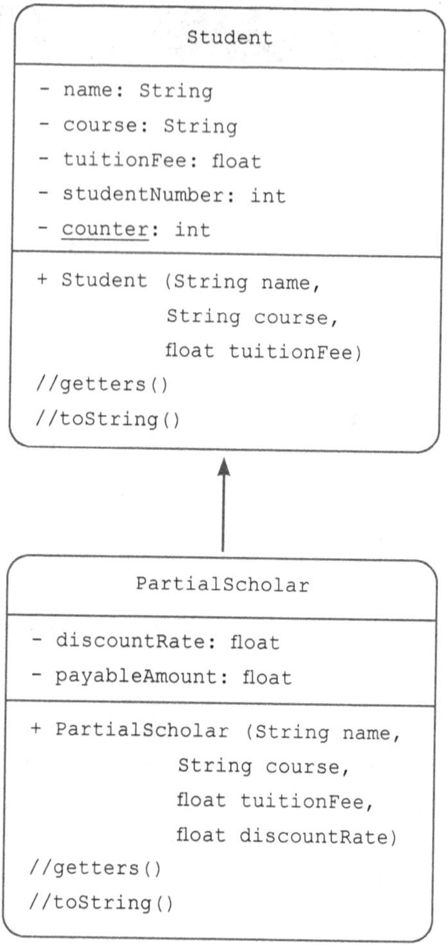

```
Student
-----------------------------------
- name: String
- course: String
- tuitionFee: float
- studentNumber: int
- counter: int
-----------------------------------
+ Student (String name,
           String course,
           float tuitionFee)
//getters()
//toString()
```

↑

```
PartialScholar
-----------------------------------
- discountRate: float
- payableAmount: float
-----------------------------------
+ PartialScholar (String name,
                  String course,
                  float tuitionFee,
                  float discountRate)
//getters()
//toString()
```

As for the `PartialScholar` class, the constructor should accept the `name` of the student, the course, the `tuitionFee` amount, and the `discountRate`. You are to reuse the constructor of the `Student` class and pass the `name`, `course`, and `tuitionFee` values. In the constructor of the `PartialScholar` class, you are to compute the `payableAmount` by using the `tuitionFee` value and deducting the product of the `tuitionFee` and the `discountRate`.

You are also required to override the `toString()` methods on both classes. Please see the sample output and the provided test class named `TestStudent`.

Here's the test class `TestStudent`, do not edit this code.

```
public class TestStudent {
        public static void main(String args[])  {
                Student students[] = {  new Student("Mathew", "BSCS", 10000.0F),
                                        new Student("Mark", "BSIT", 12000.0F),
                                        new PartialScholar("Luke", "BSIT", 12000.0F, 0.25F),
                                        new PartialScholar("John", "BSIT", 11000.0F, 0.5F),
                                };

                float totalCollection = 0.0F;

                for(int i = 0; i < students.length; i++)  {
                        System.out.println(students[i]);
                }

                // compute totalCollection value
                for(int i = 0; i < students.length; i++)  {
                        if (students[i] instanceof PartialScholar)
                                totalCollection += ((PartialScholar)students[i]).getPayableAmount();
                        else if (students[i] instanceof Student)
                                totalCollection += students[i].getTuitionFee();
                }

                System.out.println("\nTotal Collectible Amount: " + totalCollection);
        }
}
```

Here's the sample runtime output:

```
Student # 1: Mathew is taking up BSCS with a tuition fee of 10000.0

Student # 2: Mark is taking up BSIT with a tuition fee of 12000.0

Student # 3: Luke is taking up BSIT with a tuition fee of 12000.0
Discount Rate of: 25%
Net Payable Tuition Fee is: 9000.0

Student # 4: John is taking up BSIT with a tuition fee of 11000.0
Discount Rate of: 50%
Net Payable Tuition Fee is: 5500.0

Total Collectible Amount: 36500.0
```

Coding Task 2: The Animal Kingdom

Description:

You are tasked with creating a hierarchy of animal classes. The base class will be `Animal`, and you will have derived classes `Dog`, `Cat`, and `Bird`. Each animal will have a method `makeSound` which will print the sound the animal makes. You will demonstrate polymorphism by creating a method that takes an `Animal` reference and calls the `makeSound` method.

Requirements:

A. *Create the base class `Animal`:*

- It should have a method `makeSound` that prints a generic message like "`Some sound`".
- It should have a method move that prints a generic message like "`Animal is moving`".

B. *Create the derived classes `Dog`, `Cat`, and `Bird`:*

- Each class should override the makeSound method to print a specific sound, e.g., "`Bark`" for `Dog`, "`Meow`" for `Cat`, and "`Chirp`" for `Bird`.
- Each class should override the move method to print a specific movement, e.g., "`Dog is running`", "`Cat is pouncing`", and "`Bird is flying`".

C. *Demonstrate polymorphism:*

- Write a method `printAnimalSound` that takes an `Animal` reference as a parameter and calls the `makeSound` method.
- Create a list of `Animal` objects containing instances of `Dog`, `Cat`, and `Bird`.
- Iterate through the list and call the `printAnimalSound` method for each animal.
- This part of the exercise is already provided for you.

Code structure:

```
// A. Create the base class Animal:
// Base class
class Animal {
    // put codes here
```

```
}

// B. Create the subclasses Dog, Cat, and Bird:
// Subclass Dog
class Dog extends Animal {
    // put codes here
}

// Subclass class Cat
class Cat extends Animal {
    // put codes here
}

// Subclass class Bird
class Bird extends Animal {
        // put codes here
}

 // C. Demonstrate polymorphism:
 // Main class to demonstrate polymorphism
 public class AnimalKingdom {
    public static void printAnimalSound(Animal a) {
    a.makeSound();
    }

 public static void main(String[] args) {
    Animal[] animals = {  new Dog(),
                         new Cat(),
                         new Bird() };

    for (Animal animal : animals) {
      printAnimalSound(animal);
      animal.move();
    }
 }
}
```

Tasks:

- Implement the `Animal`, `Dog`, `Cat`, and `Bird` classes as described.
- Implement the `printAnimalSound` method to demonstrate polymorphism.
- Create an array of `Animal` objects and populate it with instances of `Dog`, `Cat`, and `Bird`.
- Iterate through the array, calling the `printAnimalSound` method and the move method on each animal to demonstrate polymorphism and method overriding.

Expected Output:

```
Bark
Dog is running
Meow
Cat is pouncing
Chirp
Bird is flying
```

CHAPTER 5
Other Class Features

Key Learning Objectives

- Understand and use static members, wrapper classes, and static imports.
- Apply the final keyword to variables, methods, and classes.
- Define and use enums to represent a set of predefined constants.
- Declare and implement abstract classes and methods.
- Utilize interfaces, including static and default methods.
- Implement functional interfaces and employ lambda expressions to simplify code.

This chapter explores important and specialized Java keywords and concepts that enhance code structure, functionality, and flexibility. We'll revisit the `static` keyword — understanding how static variables and methods operate at the class level — and see its role in classes like `Math` and `System`. We'll also cover wrapper classes and static imports, and delve into key object-oriented constructs such as `final`, `abstract`, `enum`, and `record`. Additionally, you'll learn how Java 8 introduced powerful functionality to interfaces — including static methods, default methods, and functional interfaces — allowing you to streamline code and implement behavior in a more flexible and reusable way. The chapter concludes by exploring how lambda expressions enable a more concise, functional style of coding, making your Java code more adaptable and elegant.

5.1 Revisiting the `static` Keyword

The `static` keyword is used in the Java Programming Language to allow you to declare a single copy of a construct (variable or method) and have that single copy be owned by the class instead of having several copies of it that are owned by each instance object.

Figure 5.1 **Diagram that shows the relationship between classes and objects**

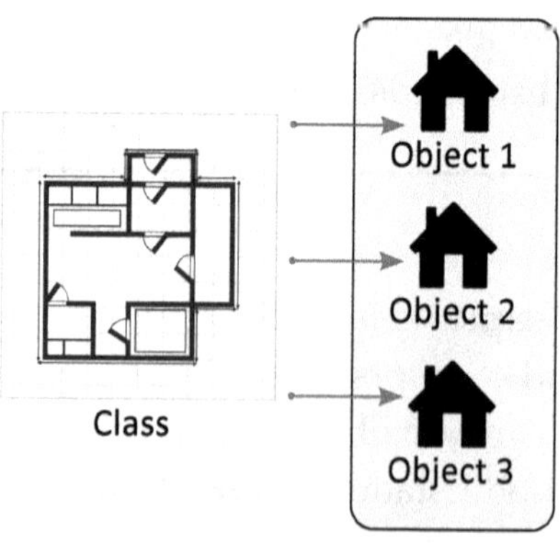

In this analogy, you can see that the class is represented by a blueprint, and the houses represent the instances of each object from the class. The green lawn can be interpreted as a static variable that is shared by all instances of the object.

The `static` keyword can only be used for attributes, methods, and inner classes. Top-level classes cannot be declared static.

Once an attribute, method, or inner class is declared as static, it means the ownership of the said construct will be owned by the class and not by any instance of any object. Even if you create several instances of a class, there will still be a single copy of a static variable.

Consider the following example:

```
// StaticExample1.java
public class StaticExample1 {

   public static int counter = 0;
   private int serialNumber;

   public StaticExample1() {

     counter++;
     serialNumber = counter;
   }

   public static void main(String[] args){

     System.out.println("Creating the first object: ");
     StaticExample1 se1 = new StaticExample1();
     System.out.println("Creating the second object: ");
     StaticExample1 se2 = new StaticExample1();
```

```
15
16      System.out.println("A total of " +
17                  counter + " objects were created.");
18    }
19 }
```

In line 2, we've declared the variable counter as `static`, and on line 3, the `serialNumber` variable is non-static. A `static` variable, also known as a class variable, will only have a single copy regardless of the number of objects that were created. Unlike a non-static variable, also known as an instance variable, it will always have the same number of copies as the number of objects created. This means that for every object instance of the class `StaticExample1`, there will exist an attribute `serialNumber` per object instance.

In lines 12 and 13, two instances of `StaticExample1` were created. Therefore, there are two copies of the instance variable `serialNumber`, one for each instance. But the `static` variable `counter` will always have a single copy because `serialNumber` is declared as `static`.

5.2 How Do We Access/Call a Static Variable or a Static Method?

We can access `static` variables/methods by using the class name instead of the instance of that class. In line 17, we have accessed `counter` directly, but if you want to access the `static` variable `counter` outside the class, you need to access it through the name of the class:

```
public class SomeOtherClass
{
    public void doSomething()
```

```
        {
                System.out.println(StaticExample1.counter);
        }
    }
```

The way we access static methods and static variables is the same, through the name of their respective classes. Unlike instance variables, both static methods and static variables are accessed through the name of the class. So if we print the serialNumber of both objects, se1 and se2, we need to call them this way:

```
// printing the serialNumber of se1
System.out.println(se1.serialNumber);
// printing the serialNumber of se2
System.out.println(se2.serialNumber);

// printing the value of counter
System.out.println(StaticExample1.counter);
```

5.3 The Math Class and the System Classes

Consider the Math class and the System class. Both of these classes belong to the java.lang package and are considered as special classes (utility classes) because all attributes and methods in both classes are static; therefore, every time you access an attribute or a method from any of these classes, you need to access them using their class names.

Check your Java documentation and examine the methods and attributes of the System and Math classes.

```
System.out.println(Math.PI);
System.out.println(Math.max(20,30));
```

In addition, the `Math` and `System` classes are the only classes that do not have any constructors in them, not even the default constructors. Neither the `Math` nor `System` classes have a `public` constructor. In fact, their constructor is `private`. The reason behind this is that all methods in both the `Math` class and the `System` classes are `static`, and therefore no object instance is needed to access the class members—both attributes and methods. They are designed so that you never make objects from them. Instead, you access their features directly through `class` name + `static` members.

5.4 The `static` Imports

This feature was introduced in Java SE 5.0 and is used to import static members (attributes and methods) individually or collectively.

Here's the syntax of static imports:

```
import static <pkg_list>.<class_name>.<member_name>;
OR
import static <pkg_list>.<class_name>.*;
```

For example, we can rewrite the previous example this way.

```
1   import static java.lang.Math.*;
2   import static java.lang.System.out;
    . . .
7   out.println(PI);
8   out.println(max(20,30));
```

In line 1, you have imported all the `static` members of the `Math` class, so every time you call a `static` member inside the `Math` class, you only need to call the variable or the method directly without involving the name of the class, just like in lines 7 and 8.

In line 2, you have imported only one `static` member inside the `System` class, and that is the `static` member `out`. This means only the `static` member `out` can be accessed without the `System` class name. So, if you want to access other `static` members like `System.gc()` or `System.exit()` in the `System` class, you still need to access them using the name of the class.

```
2   import static java.lang.System.out;
    . . .
8   out.println("This is how you print a message..");
9   System.err.println("This is an error message..");
    . . .
18  System.arraycopy(src, 0, dest, 0, src.length);
    . . .
28  System.exit(0);
```

5.5 The Wrapper Classes

Wrapper classes are the class equivalent of all Java primitive data types. They can be found inside the `java.lang` package and are used to create an object representation of a primitive value. The wrapper classes can be easily memorized. They are all spelled out completely, and the first characters are in uppercase.

Table 5.1 Wrapper class and primitive data types

Wrapper Class	Primitive Data Type
Boolean	boolean
Byte	byte
Character	char
Double	double
Float	float
Integer	int
Long	long
Short	short

5.5.1 How Do We Use the Wrapper Classes?

```
int num = 100;
Integer i = new Integer(num);          // box-in
Integer j = new Integer(200);          // box-in
int num2 = j.intValue();               // box-out
```

Box-in and box-out refer to the automatic conversion between primitive data types and their corresponding wrapper class objects. Box-in happens when a primitive type like `int`, `double`, or `char` is converted into its matching wrapper class, such as `Integer`, `Double`, or `Character`. On the other hand, box-out is when a wrapper class object is converted back to its primitive type. For example, assigning an `int` value to an `Integer` object is a box-in operation, while retrieving the primitive `int` from an `Integer` object is a box-out operation. Starting with Java 1.5, autoboxing and unboxing were introduced to handle these conversions automatically, making code easier to read and write.

In this example, we declared an `int` primitive data type with `100` as its value, and in line 8, we created an `Integer` reference variable `i` with `100` as its object value. In line 9, we also declared an `Integer` object `j` with `200` as its value.

Is this statement legal?

```
11  int sum = i + j;        // This is illegal in Java 1.4
                            // and earlier versions
```

In Java 1.4 and earlier versions, line 11 is not allowed. Because `i` and `j` are reference variables that contain reference values (or addresses) of the `Integer` objects that they represent. They do not contain the actual primitive values. Instead, they contain memory addresses of the `Integer` object.

But starting with Java SE 5 up to the present version, a feature called AutoBox-In and AutoBox-Out was introduced, which allows us to assign primitive values to Wrapper class reference variables.

The given code can be rewritten in this manner by applying autoboxing and unboxing features.

```
 7  int num = 100;
 8  Integer i = num;                    // auto boxing
 9  Integer j = 200;                    // auto boxing
10  int num2 = j;                       // unboxing
11  int sum = i + j;                    // unboxing
```

However, starting with Java 9, these constructors were deprecated, and eventually their use was strongly discouraged. The reason is simple: constructors always create a new object, even if another object with the same value already exists. This can waste memory and hurt performance.

Instead, developers are encouraged to use factory methods like `valueOf()` or rely on autoboxing. Both approaches reuse existing cached objects when possible, making the code more efficient.

The wrapper classes (`Integer`, `Double`, `Float`, `Long`, `Short`, `Byte`, `Character`, and `Boolean`) still have constructors, but they are deprecated as of Java 9. Instead of using `new Integer(10)`, always prefer `Integer.valueOf(10)` or simply write `Integer num = 10;`.

Deprecated constructors still work, but they should be avoided in new code, since they may eventually be removed in future Java versions.

Objects created from a wrapper class are commonly used to put primitive values inside a `Collection`. More discussions on the `Collection` framework in Chapter 8: Introduction to Java Collection and the Generics Framework.

5.6 The `final` Keyword

The `final` keyword can be used for attributes, methods, and classes.

Classes once declared `final` cannot be subclassed. This means that extending `final` classes will cause a compilation error. Consider the `String` class, which is declared a `final` class. This class declaration will produce a compilation error.

```
public class MyString extends String {
}
```

The `String` class under the `java.lang` package is a `final` class, and therefore, you cannot create a subclass of the `String` class.

Methods that are declared `final` cannot be overridden. Final methods can still be reused and inherited; it is just their implementation that you cannot change or override.

As for attributes, once declared as `final`, you cannot change their values. They are considered constants. Final attributes have two forms, the first is an initialized constant value. For example:

```
public static final double SALES_VAT = 0.12;
```

The `SALES_VAT` is not a variable anymore; it is now a constant value. The value assigned to `SALES_VAT` cannot be changed during the life span of your entire application; any attempt to change it will cause a compilation error.

The second type of constant declarations is the blank `final` variables. A blank `final` variable is a constant declaration that does not have any initial values assigned to it. The rules for using blank `final` variables are simple:

1. Values for blank `final` variables can be assigned only once. Once assigned, you cannot change its value.
2. A value must be assigned to a blank `final` variable within the body of the constructor. Otherwise, you cannot create an object if you have not assigned a value to a blank `final` variable.

Consider this sample code.

```
// Student.java
public class Student {

    private final int studentNumber;

    public Student() {
```

```

        studentNumber = generateStudentNumber();
    }

    private int generateStudentNumber () {

        // codes that will generate the
        // student number of each created student.
    }
}
```

At the beginning of each student's university life, they usually fill out the university application form giving their name, birthday, address, and other pertinent information. But the student number is initially left blank for the university registrar to fill in. Upon passing the entrance exam and university interview, the university registrar will then assign a student number to each admitted student. Only then will that the student applicant be a full-fledged university student with their own student number.

The student will then carry this student number during their entire stay in the university. Even after several years pass and the student graduates from the university, the student will have to give their student number in order for the university registrar to find their records.

5.7 The `record` Keyword

The `record` keyword was introduced as a 'preview feature' in Java 14 that was included as a fixed feature in Java 16. The `record` keyword is used in a class that intends to have only final variables that are set in the constructor.

Before Java 14, we could use this code:

```
// TestPoint.java
final class Point {
    public final int x;
    public final int y;

    public Point(int x, int y) {
        this.x = x;
        this.y = y;
    }

    public boolean equals(Object obj) {
    boolean result = false;

        if (obj != null && obj instanceof Point)  {
            Point p = (Point)obj;
            if (this.x == p.x && this.y == p.y)  {
                result = true;
            }
        }
        return result;
    }
    public int hashCode() {
            return x ^ y;
        }

    public String toString()   {
        return this.getClass().getName() +
            "[x=" + this.x + ", y=" + this.y + "]";
    }
}
```

```
public class TestPoint
{
  public static void main(String[] args)
  {
      Point p1 = new Point(0, 0);
      Point p2 = new Point(10, 20);
      Point p3 = new Point(10, 20);

      System.out.println("Point 1 is at: "
               + p1.x + " and " + p1.y);
      System.out.println("Point 2 is at: "
               + p2.x + " and " + p2.y);
      System.out.println("Point 3 is at: "
               + p3.x + " and " + p3.y);

      System.out.println(p1);
      System.out.println(p2.toString());
      System.out.println(p3);

      System.out.println("Comparing p1 and p2: "
               + p1.equals(p2));
      System.out.println("Comparing p2 and p3: "
               + p2.equals(p3));
  }
}
```

This pre-Java 14 code will compile and run with the following output:

```
Point 1 is at: 0 and 0
Point 2 is at: 10 and 20
Point 3 is at: 10 and 20
Point[x=0, y=0]
Point[x=10, y=20]
Point[x=10, y=20]
Comparing p1 and p2: false
Comparing p2 and p3: true
```

Since `Point.java` is a user-defined class, the `equals()`, `hashCode()`, and `toString()` methods need to be overridden from the `Object` class.

Java 14 onwards, we can write the same code using the `record` preview feature, which implicitly overrides the implementation of the `equals()`, `hashCode()`, and `toString()` methods from the `Object` class. This version is shorter and easier to write.

```
// TestPoint.java
record Point(int x, int y) { }

public class TestPoint
{
 public static void main(String[] args)
  {
     Point p1 = new Point(0, 0);
     Point p2 = new Point(10, 20);
     Point p3 = new Point(10, 20);

     System.out.println("Point 1 is at: "
              + p1.x() + " and " + p1.y());
```

```
        System.out.println("Point 2 is at: "
                  + p2.x() + " and " + p2.y());
        System.out.println("Point 3 is at: "
                  + p3.x() + " and " + p3.y());

        System.out.println(p1);
        System.out.println(p2.toString());
        System.out.println(p3);

        System.out.println("Comparing p1 and p2: "
                  + p1.equals(p2));
        System.out.println("Comparing p2 and p3: "
                  + p2.equals(p3));

    }
}
```

This version can be compiled using the `--enable-preview` and the `-source` options. To compile using Java 14:

```
javac --enable-preview -source 14 TestPoint.java
```

After a successful compilation, we execute the program by using Java 14 and a virtual machine option of `--enable-preview` during execution. To run:

```
java -enable-preview TestPoint
```

When executed, this is the expected output:

```
Point 1 is at: 0 and 0
Point 2 is at: 10 and 20
```

```
Point 3 is at: 10 and 20
Point[x=0, y=0]
Point[x=10, y=20]
Point[x=10, y=20]
Comparing p1 and p2: false
Comparing p2 and p3: true
```

Notice that the Java 14 code is way shorter than the pre-Java 14 code. The output, however, is the same.

To compile using Java 16 and newer versions, you can compile it directly:

```
javac TestPoint.java
```

5.8 The enum Keyword

Before Java 1.5 came out, the concept of enumeration already existed, but Java had another way of handling it. The `enum` keyword is used to define a special data type called an enumeration, which represents a fixed set of constants, such as days of the week or directions.

```
public class InnerPlanets {
  public static final int MERCURY       = 0;
  public static final int VENUS         = 1;
  public static final int EARTH         = 2;
  public static final int MARS          = 3;

  public static void main(String args[]) {
    for (int i = 0; i < 4; i++) {
      switch(i) {
```

```
            case MERCURY: System.out.println("Mercury");
                          break;
            case VENUS  : System.out.println("Venus");
                          break;
            case EARTH  : System.out.println("Earth");
                          break;
            case MARS   : System.out.println("Mars");
                          break;
            default     : System.out.println("Other planets");
        }
      }
    }
}
```

Starting with Java 1.5, `enum` was added as a new language feature that allows the declaration of a fixed set of constant values. Some examples where you can use enums are as follows: the days of a week, the number of hours per day, the months of the year, or even the suits and ranks of a set of playing cards. These examples will always have a fixed set of values.

All the values declared inside an `enum` are constants and, therefore, usually written in all uppercase. It also allows a set of values to be declared in a descriptive manner.

Here is a sample code demonstrating the use of `enum`.

```
// InnerPlanets.java
public enum InnerPlanets {
    MERCURY, VENUS, EARTH, MARS;
}
```

```
// TestInnerPlanets.java
```

```
public class TestInnerPlanets {
  public static void main(String args[]) {
    for(InnerPlanets planets:InnerPlanets.values()){
      switch(planets) {
        case MERCURY : System.out.println("Mercury");
                       break;
        case VENUS   : System.out.println("Venus");
                       break;
        case EARTH   : System.out.println("Earth");
                       break;
        case MARS    : System.out.println("Mars");
      }
    }
  }
}
```

In the old enumeration, there is no type-safety feature checking done at compile time. But with this Java 5 feature, it limits the values of a set of `enum` values that are checked at compile time, thus eliminating the need for the `default` statement in your `switch-case`.

The enumerated types can also be declared inside a class body, like in this example:

```
// TestInnerPlanets.java
public class TestInnerPlanets  {
  enum InnerPlanets   {
       MERCURY, VENUS, EARTH, MARS;
  }

  public static void main(String args[]) {
```

```
8          System.out.println("Our home planet: "
9                                  + InnerPlanets.EARTH);
10     }
11 }
```

Since enumerated types are like classes, the rules on access modifiers apply to both top-level classes and top-level enum types. Both can only have either a public or a default access modifier.

If an enum type is declared inside another class, it can have any access modifier assigned to it. Also, all values declared inside an enum type are static and final values and should be accessed through the name of the enum type, like InnerPlanets.EARTH in lines 8 and 9.

Here are some valid enum declarations.

```
enum TrafficLight { RED, YELLOW, GREEN }
enum Day { MONDAY, TUESDAY, WEDNESDAY,
               THURSDAY, FRIDAY, SATURDAY, SUNDAY }
enum Month {JANUARY, FEBRUARY, MARCH,
               APRIL, MAY, JUNE
               JULY, AUGUST, SEPTEMBER,
               OCTOBER, NOVEMBER, DECEMBER}
enum Direction { NORTH, SOUTH, EAST, WEST }
```

5.9 The abstract Keyword

The abstract keyword can only be used for classes and methods. The abstract keyword means "there is no concrete implementation"; therefore, abstract methods do not have any implementation in their method body, just the method signature. As a rule, if a class contains an `abstract` method, the class must also be declared as `abstract` itself.

If a class is declared as `abstract`, you cannot instantiate objects from an `abstract` class. Abstract classes serve as a template that forces their subclasses to override methods that should be present in all animal subclasses. In our example, the `abstract` class `Animal` contains both `abstract` and non-abstract methods that must be present in all types of animal objects. All animals eat, breathe, and sleep, but the implementation of these methods differs from each type of animal. Therefore, we have declared eat, breathe, and sleep methods as `abstract, forcing` all subclasses of the `Animal` class to have a concrete implementation of these methods. The `play` method, on the other hand, is a non-abstract method and has a concrete implementation, also known as a method body. Overriding the method `play` is optional and is not required because it is a non-abstract method.

Figure 5.2 **Abstract class `Animal` diagram**

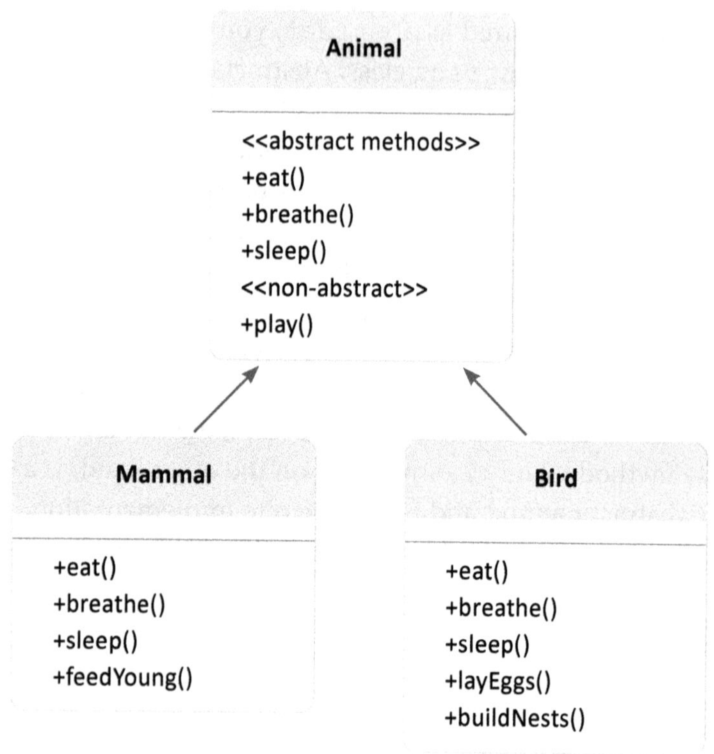

Consider the given UML. The `Animal` class has `abstract` methods. Therefore, all subclasses of the `Animal` class MUST override all `abstract` methods of the superclass `Animal`.

Here is the Java code equivalent of the given UML above.

```
// Animal.java
public abstract class Animal {
 // abstract methods
 public abstract void eat();
 public abstract void breathe();
```

```
6   public abstract void sleep();
7
8   // non-abstract methods
9   public void play() {
10         System.out.println("Have fun.");
11   }
12 }
```

Lines 4, 5, and 6 are `abstract` method declarations. They do not have any body in them, and they end with a semicolon (`;`).

The `abstract` class `Animal` serves as a template for all types of animal objects. If you declare a class of type `Animal` (a subclass of class `Animal`), that class must override all `abstract` methods of class `Animal`, or else that class will end up being an `abstract` class itself because it will inherit all `abstract` methods.

Here's the class `Mammal` and the class `Bird`.

```
// Mammal.java
public class Mammal extends Animal {
   // override all abstract methods
   public void eat() {
       System.out.println("Mammal Eats");
   }
   public void breathe() {
       System.out.println("Mammal Breathe");
   }
   public void sleep() {

       System.out.println("Mammal Sleeps");
   }
```

```

    // Mammal only methods
    public void feedYoung() {
        System.out.println("Feed Young");
    }
}
```

```
// Bird.java
public class Bird extends Animal {
    // override all abstract methods
    public void eat() {
        System.out.println("Bird Eats");
    }
    public void breathe() {
        System.out.println("Bird Breathe");
    }
    public void sleep() {
        System.out.println("Bird Sleeps");
    }

    // my own birdly methods
    public void layEggs() {
        System.out.println("Lay Eggs");
    }
    public void buildNest() {
        System.out.println("Build Nests");
    }
}
```

Here is our test class.

```
// TestAnimal1.java
public class TestAnimal1 {
    public static void main(String[] args) {
        // cannot instantiate an object
        // from an abstract class
        //! Animal a = new Animal();

        Mammal dog = new Mammal();
        Bird eagle = new Bird();

        doAnimalStuff(dog);
        doAnimalStuff(eagle);
    }
    public static void doAnimalStuff(Animal a) {
        if (a instanceof Mammal m) {

                m.eat();                // redundancy
                m.breathe();            // redundancy
                m.sleep();              // redundancy
                m.play();               // redundancy
                m.feedYoung();
        }
        else if (a instanceof Bird b) {

                b.eat();                // redundancy
                b.breathe();            // redundancy
                b.sleep();              // redundancy
                b.play();               // redundancy
                b.layEggs();
                b.buildNest();
        }
    }
}
```

But this test class is not that well written; there are a lot of redundancies. Lines 17 to 20 and lines 25 to 28 are redundant code. To remove redundancies in our code, let us refactor the code by moving the repeated method calls to a shared superclass. For instance, since `Mammal` and `Bird` both extend the `Animal` class and `Animal` has methods like `eat()`, `breathe()`, `sleep()`, and `play()`, you can invoke these directly on the `Animal` reference, removing the need for type-specific casts.

Consider the revised test class:

```
// TestAnimal2.java
public class TestAnimal2 {
    public static void main(String[] args) {
        Mammal dog = new Mammal();
        Bird eagle = new Bird();

        doAnimalStuff(dog);
        doAnimalStuff(eagle);
    }
    public static void doAnimalStuff(Animal a) {
        a.eat();
        a.breathe()
        a.sleep();
        a.play();

        if (a instanceof Mammal m) {

                m.feedYoung();
        }
        else if (a instanceof Bird b) {

                b.layEggs();
                b.buildNest();
        }
    }
}
```

Now there are no more redundancies, and the code is cleaner. Lines 11 to 14 that call the methods from both `Bird` and `Mammal` were now called once using the `Animal a` reference variable.

5.10 Java Interfaces

In Java, interfaces are used to define a contract or a set of methods that a class must implement. They allow you to specify behaviors that different classes can share, even if the classes don't share a common parent. This makes it possible for unrelated classes to provide consistent functionality, ensuring code flexibility and modular design. By using interfaces, developers can write code that is more reusable and maintainable, as it separates the "what" from the "how," focusing on defining actions without dictating the specific implementation details. Interfaces allow you to specify methods that a class must implement without dictating how those methods are executed, enabling different classes to provide their own specific implementations. All methods in the interface are abstract and should be overridden by the implementing class, and like an abstract class, we cannot instantiate an object using the keyword `new` because an interface is not a "concrete" class. This is particularly useful for designing loosely coupled and modular systems, as interfaces define what a class can do rather than how it does it.

Figure 5.3 The Java interface diagram

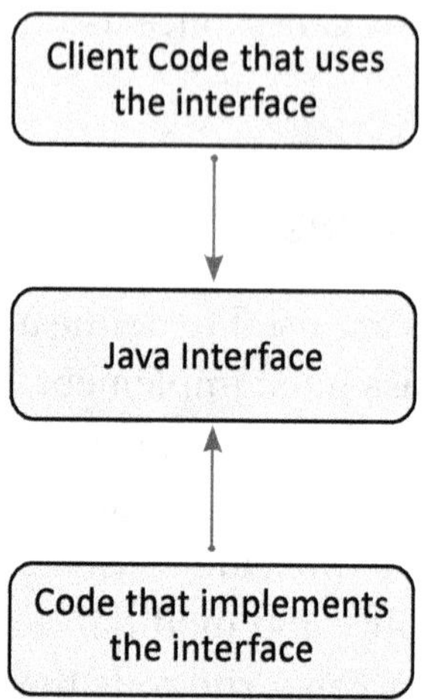

An interface can also have variables inside it, but once a variable is declared, it is automatically converted into a `public static final` variable. These variables are now treated as constants, which means we cannot change the values of the variables in an interface. All classes that implement the interface will inherit these constant values. Consider the set of interfaces:

```
interface I1 { double pi = 3.14; }
interface I2 { public double pi = 3.14; }
interface I3 { static double pi = 3.14; }
interface I4 { final double pi = 3.14; }
interface I5 { public static double pi = 3.14; }
interface I6 { static final double pi = 3.14; }
interface I7 { public final double pi = 3.14; }
```

All the interfaces declared a variable `pi`, which translates to:

```
public static final double pi = 3.14;
```

Any class that will inherit `pi` will not be able to change its value because it is inherently a constant.

To declare an interface, we use the following syntax:

```
<modifier> interface <name> [extends
                                <superInterface>*]
{
      <member_variables>*
      <member_methods>*
}
```

Take note that interfaces cannot have constructors. It would be senseless to have a constructor because we cannot create an object from an interface. An interface can also `extend` other interfaces. In an event that there are multiple interfaces that need to be extended, a comma is needed to separate the inherited interfaces.

```
interface I1 { public void doStuff(); }
interface I2 { public void doSomething(); }
interface I3 extends I1, I2 {
   public void doSomethingElse();
}
```

Figure 5.4 The `FlyingObject` interface

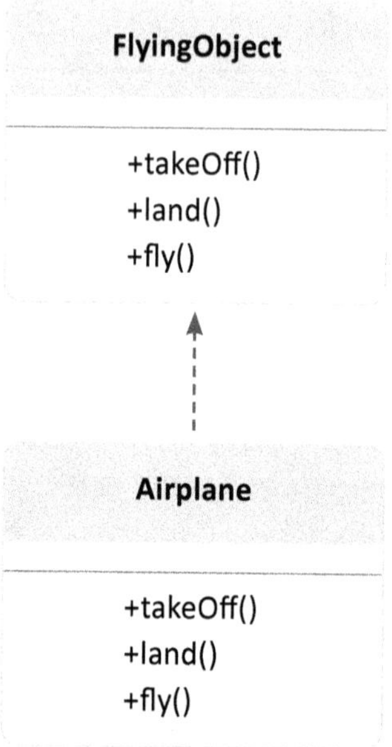

The `FlyingObject` interface is implemented by an `Airplane` class that overrides the `takeOff()`, `land()`, and `fly()` methods of the interface.

The syntax for using an interface is as follows:

```
<modifier> class <name> [extends <superclass>]
                 [implements <interface>
                 [,<interface>]*]
{
      <member_declaration>*
}
```

```
// FlyingObject Interface: FlyingObject.java
      public interface FlyingObject {

          public void takeOff();
          public void land();
```

```
        public void fly();
    }
```

```
// Airplane class that implements
// FlyingObject: Airplane.java

public class Airplane implements FlyingObject {

    public void takeOff() {

        // increase speed
        // lift landing gear
    }
    public void fly() {

        // keep those engines running
    }
    public void land() {

        // lower flaps
        // reduce speed
    }
}
```

An interface can also be used to simulate multiple inheritance by allowing a class to implement multiple interfaces or to implement an interface while extending a class.

Consider the given diagram below, the class `Bird` extends the `Animal` class while implementing the `FlyingObject` interface. Since all methods in the interface `FlyingObject` are abstract classes, the `Bird` class needs to override the methods `takeOff()`, `land()`, and `fly()`. The same goes for the `abstract` class `Animal`, the `Bird` class should also override the abstract methods `eat()`, `breathe()`, and `sleep()`. As for the `play()` method, overriding it is optional because it is a concrete method and not an `abstract` method. On top of that, the class `Bird` can also declare its own unique set of methods exclusively found in a `Bird` object, like `feedYoung()` and `buildNest()`.

Figure 5.5 The *FlyingObject* interface used to apply multiple inheritance in Java

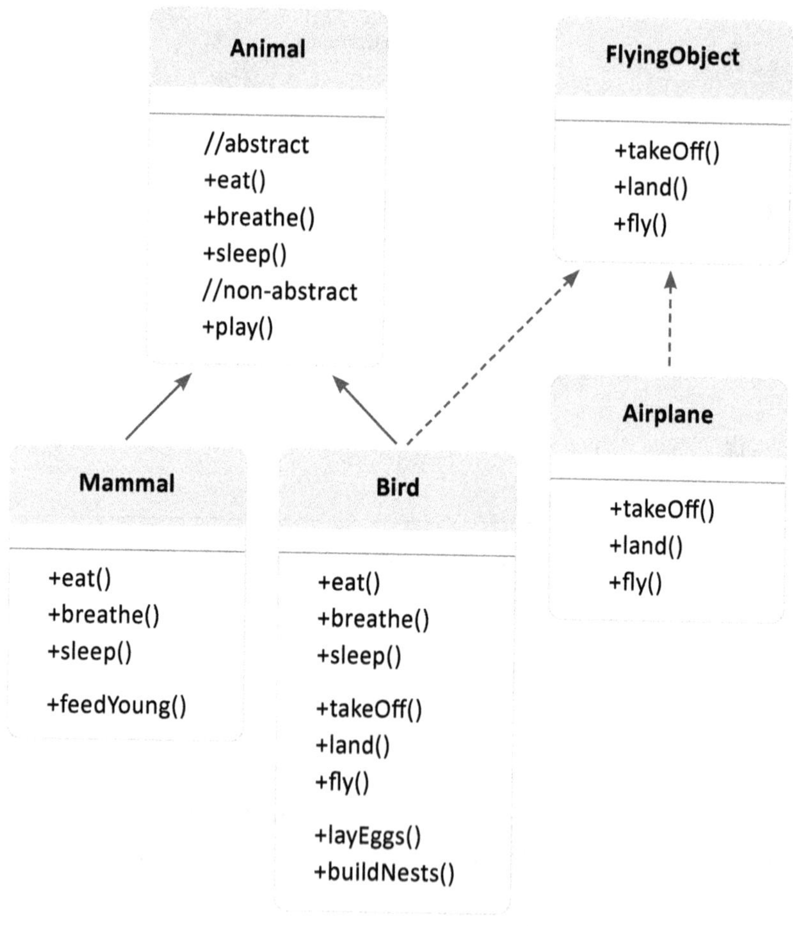

The `Bird` class will now have this code:

```
// Bird.java
public class Bird extends Animal
                        implements FlyingObject {
  // override all abstract methods
  public void eat() {

        System.out.println("Bird Eats");
  }
  public void breathe() {

        System.out.println("Bird Breathes");
  }
  public void sleep() {

        System.out.println("Bird Sleeps");
  }

  // my own birdly methods
  public void layEggs() {

        System.out.println("Lay Eggs");
  }
  public void buildNest() {

        System.out.println("Build Nests");
  }
  // overriding all methods
  // from FlyingObject interface
  public void takeOff() {

```

```
        // flap birdy wings
    }
    public void fly() {
        // flap...fly...soar
    }
    public void land() {
        // ready to land little birdy
    }
}
```

In Java 8, a new set of language features was included. Interfaces in Java 8 can now have `default` methods and `static` methods. Also, the concept of *FunctionalInterfaces* was introduced, as well as the lambda (`->`) operator.

5.11 The Interface `default` Methods

Before Java 8, all methods in an interface were abstract methods. This means that once a `class` implements an `interface`, all the methods of that interface should be overridden by the implementing `class`. Since Java 8, the `default` method can now be included in an interface.

This feature allows developers to create a scalable interface, an interface that can be updated without causing an error to the implementing classes.

Consider this pre-Java 8 code:

```
  // Animal.java
public interface Animal  {
    public void eat();
    public void sleep();
    public void breathe();
}
```

```
// Mammal.java
public class Mammal implements Animal  {
    @Override
    public void eat()  {
        // some codes here
    }
    @Override
    public void sleep(){
        // some codes here
    }
    @Override
    public void breathe(){
        // some codes here
    }
}
```

This code will compile, and it looks good for now, but what if we update the `Animal interface` by adding another method to it? Then the `Mammal class` will have a compilation error complaining that the newly added method in the `interface Animal` was not overridden.

To address the problem of scalability, a `default` method can be added in the `Animal` interface so that the implementing `Mammal` class will not have any compilation error after the updated `Animal` interface is compiled.

```
// Animal.java
public interface Animal  {
  public void eat();
  public void sleep();
  public void breathe();
  public default void play()  {
      System.out.println("Animal is having fun.");
  }
}
```

```
// Mammal.java
public class Mammal implements Animal  {
  @Override
  public void eat()  {
      // some codes here
  }
  @Override
  public void sleep(){
      // some codes here
  }
  @Override
  public void breathe(){
      // some codes here
  }
}
```

A `default` method in an interface must explicitly be declared `default`, not to be confused with the `default` statement in a `switch-case` construct or even the *default* (package-friendly) access modifiers. Default methods are implicitly `public` methods and can be overridden by the implementing class. If the implementing class chooses not to override the `default` method, then the implementation of the `default` method will execute once it is called.

A `default` method is only allowed to be either `public` or have a *default* access modifier. The `protected` access modifier is not allowed to be used.

If the `default` method is overridden in the implementing class, it can still call the `default` method in the interface by using the syntax for calling the `default` method in the super interface: *`InterfaceName.super.defaultMethod();`*.

Consider the example below:

```
// DefaultInterface.java
interface MyInterface  {
 public void sayHi();

 default void myDefaultGreeting()  {
   System.out.println("Hello default world.");
 }
}
public class DefaultInterface
             implements MyInterface  {
 @Override
 public void sayHi(){
   System.out.println("Hi");
 }

 @Override
 public void myDefaultGreeting() {
   MyInterface.super.myDefaultGreeting();
   System.out.println("Overriding default method.");
 }
 public static void main(String args[]) {
   DefaultInterface di = new DefaultInterface();
   di.sayHi();
   di.myDefaultGreeting();
 }
}
```

This should give us a runtime output of:

```
Hi
Hello default world.
Overriding the default method.
```

If this code causes a compilation error, you may need to recheck your installed JDK version. This feature will not work in pre-Java 8 versions. If you have Java 8 or newer Java versions, then this code should compile.

The `default` methods in an `interface` solve the scalability issue, enabling the `interface` to declare new methods while being implemented by existing classes. However, the `default` methods will not force their implementation on the existing classes that use that interface.

5.12 The Use of Default Methods in Implementing Multiple Interfaces

One of the advantages of interfaces in Java is its ability to simulate multiple inheritance without the drawbacks of implementing a method that is present in both parent classes.

If two identical `default` methods are present in two interfaces that are implemented by the same class, the *`InterfaceName.super.defaultMethod();`* syntax can be used in this scenario.

```
interface MyInterface1  {
 public default void myDefaultGreeting()  {
     System.out.println("Inside MyInterface 1");
 }
}
```

```
interface MyInterface2  {
  public default void myDefaultGreeting()  {
        System.out.println("Inside MyInterface 2");
  }
}

public class DefaultInterfaceMultipleInheritance
        implements MyInterface1, MyInterface2 {
  // overriding default method myDefaultGreeting()
  @Override
  public void myDefaultGreeting() {
    // calling the method in MyInterface1
    MyInterface1.super.myDefaultGreeting();

    // calling the method in MyInterface2
    MyInterface2.super.myDefaultGreeting();

    System.out.println("Overriding the " +
                           "default method.");
  }
  public static void main(String args[])  {
    DefaultInterfaceMultipleInheritance di =
      new DefaultInterfaceMultipleInheritance();
    di.myDefaultGreeting();
  }
}
```

When executed, we will get this output:

```
Inside MyInterface 1
Inside MyInterface 2
Overriding the default method.
```

5.13 The Interface `static` Methods

Since Java 8, interfaces can also declare static methods, in addition to `public` and `abstract methods`. These are methods declared inside an interface that have an actual method body and are declared static. Since these are static methods, they can be called through the name of the interface. Here is an example of an interface static method:

```
// MyInterface.java
public interface MyInterface  {
  // a public abstract method
  public void doStuff();

  // a static method
  public static void doSomething()  {
      System.out.println("Inside doSomething..");
  }
}
```

To test the given static method, here is an example of a test class:

```
public class StaticInterface
            implements MyInterface  {

  @Override
  public void doStuff()  {
      System.out.println("Inside doStuff..");
  }
  public static void main(String[] args)  {
      StaticInterface si = new StaticInterface();
      si.doStuff();
      MyInterface.doSomething();
  }
}
```

The output will be:

```
Inside doStuff..
Inside doSomething..
```

The static methods are used to create utility or helper methods to be declared inside an interface. It is a part of the interface where it is declared and can only be used by calling the method through the interface name (ie, `MyInterface.doSomething()`).

5.14 Private Methods in Interfaces (JDK 9)

Private methods in interfaces allow you to share code between `default` and `static` methods inside the same `interface`, without exposing those helper methods to the implementing classes.

Here's an example:

```
// MyInterface.java
public interface MyInterface {

  // A public abstract method
  void doStuff();

  // A static method
  static void doSomething() {
      System.out.println("Inside doSomething..");
      logCommon(); // calling private helper method
  }

  // A default method
  default void doMore() {
      System.out.println("Inside doMore..");
```

```
            logCommon(); // calling private helper method
    }

    // A private helper method (JDK 9+)
    private static void logCommon() {
        System.out.println("[LOG] This is a private "
                    + "helper inside the interface.");
    }
}
```

```
// StaticInterface.java
public class StaticInterface
                implements MyInterface {
  @Override
  public void doStuff() {
      System.out.println("Inside doStuff..");
  }

  public static void main(String[] args) {
      StaticInterface si = new StaticInterface();
      si.doStuff();
      si.doMore();      // calls default method
                        // in the interface

      MyInterface.doSomething(); // calls static
                        // method in the interface
  }
}
```

In this code, the interface `MyInterface` demonstrates how `private` methods can be used to support code reuse inside `interfaces` (a feature added in JDK 9). The interface defines three kinds of methods: an `abstract` method (`doStuff()`), a `static` method (`doSomething()`), and a `default` method (`doMore()`). Both the `static` and `default` methods call the `private` helper method `logCommon()`, which prints a log message. This `private` method is not visible to the implementing class (`StaticInterface`), but it allows common logic to be shared between `doSomething()` and `doMore()` without repeating code. In the `main` method, an object of `StaticInterface` is created. Calling `si.doStuff()` runs the class's own implementation, while `si.doMore()` executes the `default` method defined in the `interface`. Meanwhile, `MyInterface.doSomething()` runs the interface's `static` method. Both the `default` and `static` methods successfully reuse the hidden logging code from the `private` method, showing how `private` methods in `interfaces` improve code organization and maintainability.

5.15 Sealed Classes (JDK 17+)

Starting in JDK 17, Java supports both `sealed` classes (and interfaces) and pattern matching for `instanceof` as standard features. When these two features are used together, your code becomes:

- More readable
- Safer (thanks to exhaustive type checking)
- Cleaner, often eliminating the need for a final else clause

This combination improves how we model limited hierarchies, especially in domain-specific designs like shapes, operations, or events.

What Are Sealed Classes?

A sealed class or interface allows you to strictly define which other classes can extend or implement it.

```
sealed interface Shape permits Circle, Rectangle {}
```

This tells the compiler that only `Circle` and `Rectangle` are allowed to implement the `Shape` interface. No other class can implement `Shape`, which helps the compiler reason better about the program.

Here's the full code:

```
// Main.java
sealed interface Shapes
      permits Circle, Rectangle {}

final class Circle implements Shapes {
   int radius = 3;
}

final class Rectangle implements Shapes {
   int width = 4, height = 6;
}

public class Main {
   public static void describeShape(Shapes shape){
        if (shape instanceof Circle c) {
            System.out.println("Circle with radius "
                             + c.radius);
       } else if (shape instanceof Rectangle r) {
       System.out.println("Rectangle of "
           + r.width + " x " + r.height);
       }
 }
```

```
    public static void main(String[] args) {
        describeShape(new Circle());
        describeShape(new Rectangle());
    }
}
```

Here's the code breakdown:

1. Sealed Interface Definition

```
sealed interface Shape permits Circle, Rectangle {}
```

This restricts the `Shape` hierarchy to only two known types: `Circle` and `Rectangle`.

2. Concrete Classes

```
final class Circle implements Shape {
int radius = 3;
}
final class Rectangle implements Shape {
int width = 4, height = 6;
}
```

These are the only permitted implementers of `Shape`, and they are marked `final`, which means they can't be subclassed further.

3. Pattern Matching in Action

```
if (shape instanceof Circle c) { ... }
else if (shape instanceof Rectangle r) { ... }
```

Each instanceof condition not only checks the type but also binds the matched object to a variable (`c` or `r`) for use in the block. No need to cast manually – the compiler handles it.

4. No `else` Clause Needed

Because the compiler knows all possible subtypes of `Shape`, it can verify exhaustiveness. This means if any subtype is missing in the logic, the compiler can warn you – much like how `switch` on `enums` works.

Why Is This Useful?

- Compile-time safety: The compiler knows all allowed subtypes, so it can catch missing cases.
- Better readability: Fewer casts and clearer control flow.
- Improved maintainability: Adding a new permitted subclass will cause a compilation error in the pattern match until it's handled.

All permitted subclasses must be `final`, or `sealed`

5.16 The Functional Interface and the Lambda (->) Operator

A functional interface is essentially an interface that has only one abstract method declared. On top of the interface declaration, an `@FunctionalInterface` annotation is placed to indicate the interface is intended to be functional.

An annotation is a form of metadata that provides additional information about the code without affecting its behavior. Annotations can be used to give instructions to the compiler, define configurations, or assist during runtime processing. In this example, the `@FunctionalInterface` annotation tells the compiler that this interface contains only one method, which makes it a `FunctionalInterface`.

```
@FunctionalInterface
interface Greeter {
    public void sayHello(String name);
}
```

What is important is that the functional interface should only have one abstract method, and it can also have any number of default methods. Functional Interfaces can also be found in the standard API, like the `Runnable`, `Comparable`, and `ItemListener interfaces`. These interfaces only have one abstract method, and therefore, they are considered functional interfaces.

Using the lambda (->) operator, we can easily provide an implementation of the methods without creating any additional classes. Consider this sample code:

```
// LambdaExample.java
@FunctionalInterface
interface Greeter {
 public void sayHello(String name);
}

public class LambdaExample {
 public static void main(String args[])  {
     Greeter g = (name) ->
     System.out.println("Hello " +
                          name + "!");
     g.sayHello("Lawrence");
  }
}
```

This example will pass the value "Lawrence" to the method sayHello and display it.

We can also have a lambda (->) operator to point to an entire code block. If there are two or more statements that need to be executed, then an enclosing pair of curly braces ({ }) is needed.

```
Greeter g = (name) -> {
            System.out.print("Hello ");
            System.out.println(name);
            };
g.sayHello("Lawrence");
```

A method with multiple arguments and a return type can be called by a lambda (->) operator.

```
// LambdaAnotherExample.java
@FunctionalInterface
interface Greeter {
  public String afterBirthday(String name,
                 int age);}

public class LambdaAnotherExample {
 public static void main(String args[])  {
    Greeter g = (name, age) -> {
          return name + ", next year you will be "
          + (age + 1) + " years old.";
    };
    System.out.println(g.afterBirthday("Joe", 30));
 }
}
```

This will give us an output of:

```
Joe, next year you will be 31 years old.
```

Chapter Summary

- In Java, the `static` keyword is used to define class-level variables and methods that can be accessed without creating an instance of the class. Static members are shared among all instances of the class, meaning they belong to the class itself rather than any specific object.
- Both the `Math` and `System` classes in the `java.lang` package contains `static` attributes and `static` methods. In both classes, there are constructors because there's no need for us to create an object in these methods, as they are designed to only have a single copy (or single instance) of their respective attributes and methods.
- The `static` imports allow for the import of `static` members (fields and methods) from other classes, enabling them to be used without specifying the class name. This feature simplifies code by allowing direct access to static members, improving readability and reducing verbosity.
- Wrapper classes in Java provide a way to use primitive data types (like `int`, `char`, etc.) as objects, enabling them to be used in contexts that require objects, such as collections. They also offer utility methods for converting between primitives and their corresponding wrapper objects, as well as useful constants and methods for handling data.
- The `final` keyword in Java is used to declare constants, prevent method overriding, and restrict inheritance. When applied to variables, it makes them immutable; when used with methods, it prevents them from being overridden; and when used with classes, it ensures they cannot be subclassed.
- The `record` keyword in Java is used to define a special kind of class that automatically generates boilerplate code

for immutable data carriers, such as constructors, getters, `equals()`, `hashCode()`, and `toString()` methods. Records simplify the creation of classes that are primarily used to store data by reducing verbosity and improving code readability.

- The `enum` keyword in Java is used to define a set of named constants, representing a fixed collection of values. Enums provide type safety, making code more readable and less error-prone by ensuring that variables can only hold predefined constant values.
- The `abstract` keyword in Java is used to define abstract classes and methods that cannot be instantiated or directly implemented. Abstract classes serve as a blueprint for subclasses, while `abstract` methods must be overridden by concrete subclasses, enforcing a contract for specific behavior.
- In Java, the `interface` keyword is used to declare a contract for classes to implement certain methods without specifying how they should be implemented. Interfaces allow the application of multiple inheritance concepts in Java, enabling classes to adhere to common behaviors across unrelated class hierarchies. They promote code reusability, flexibility, and maintainability by defining a clear separation between API and implementation details.
- In Java 8, the introduction of `default` methods in interfaces allows method implementations to be included directly within the interface, providing a way to add new functionality without breaking existing implementations. Default methods enable interfaces to evolve over time while maintaining backward compatibility, as classes implementing the interface can either inherit the default implementation or override it with their own. This feature promotes greater flexibility and reusability in code design.

- Another language feature update in Java 8, interfaces can now contain static methods, which are utility methods that belong to the interface itself rather than any implementing class. These static methods can be called directly on the interface, providing common functionality related to the interface without requiring an instance of any implementing class.
- A Functional Interface is an interface that contains exactly one abstract method. They are used to represent functions as objects, allowing them to be passed around as method parameters or returned from methods. The Lambda (->) Operator in Java is used to define anonymous functions quickly and concisely, making it easier to implement Functional Interfaces without the need for verbose anonymous inner classes. This combination enables more expressive and functional-style programming in Java, promoting cleaner and more readable code.

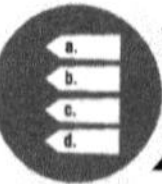

Multiple Choice Questions

1. **Given the code snippet:**

```
class MyString extends String  {
 // some codes here
}
```

The `String` class is a final class. Which of the following statements below is NOT TRUE with regard to the code snippet?

a. This code will compile and will inherit all the content of the `String` class.

b. The class `MyString` can extend any class as long as it is NOT a final class.

c. The class `String` is a final class and therefore cannot be inherited.

d. The class `MyString` will cause a compilation error.

2. **Which of the following statements is NOT TRUE about abstract classes?**

a. You can declare a top-level final abstract class.

b. You can declare an abstract class with only abstract methods inside it.

c. Abstract classes cannot be instantiated.

d. You can declare an abstract class with a regular, non-abstract method inside it.

3. **What is the correct way to declare an `abstract` method `eat()` that does not return anything, and can be accessed anywhere?**
   a. `public abstract void eat();`
   b. `public abstract Object eat() { return null; }`
   c. `public abstract void eat() { }`
   d. `public abstract void eat() { return null; }`

4. **Given the code of an interface:**

```
public interface MyInterface  {
      public abstract void doSomething();       // 1
      public void doOtherThing()  { }            // 2
      public int returnAnIntMethod();            // 3
      public void doNothing();                   // 4
}
```

Which lines of the code will cause a compilation error?

a. Line 1 will cause a compilation error.
b. Line 2 will cause a compilation error.
c. Line 3 will cause a compilation error.
d. Line 4 will cause a compilation error.
e. No error in the whole code.

5. **Given the code below:**

```
interface Flyable {}
class Bird implements Flyable {}
class PetBird extends Bird {}
```

Which of the following declarations will cause a compilation error?

a. `Bird bb = new Bird();`
b. `Bird bf = new Flyable();`
c. `PetBird pp = new PetBird();`
d. `Flyable fp = new PetBird();`

6. **Following Java naming conventions, which would best declare a constant String birthday?**

a. `final String birth_day = "June 1, 2000";`
b. `final String birthDay = "June 1, 2000";`
c. `final String BIRTHDAY = "June 1, 2000";`
d. `final String BirthDay = "June 1, 2000";`

7. **What do you call an interface that should have only one method in it?**

a. Singleton Interface
b. Mono Interface
c. Singular Interface
d. Functional Interface

8. **Given the following constants below, which one is NOT fit to be declared as an Enumeration (`enum`)?**

a. Months of the Year: January to December
b. Playing Cards: Diamonds, Hearts, Spades, Clubs
c. Days of the Week: Monday to Sunday
d. Name of your friends: Lawrence, Percy, Jerry, and Ericson

9. **Given the code below that declares all the different sizes of coffee, how will you assign the value to the variable `size`?**

```
enum CoffeeSize   {
     SHORT, TALL, GRANDE, VENTI
};
class Coffee   {
     CoffeeSize size;
}
public class CoffeeTest1   {
     public static void main(String args[])   {
          Coffee drink = new Coffee();
          // insert code here
          System.out.println(drink.size);
     }
}
```

Which of the following codes, when inserted in the comment inside the main method, will cause the code to run properly?

a. `drink.size = SHORT;`

b. `drink.size = CoffeeSize.0;`

c. `drink.size = Coffee.SHORT;`

d. `drink.size = CoffeeSize.SHORT;`

10. Which of the following IS NOT a Wrapper Class?

- **a.** `Boolean`
- **b.** `Character`
- **c.** `String`
- **d.** `Integer`

Multiple Choice Answer Key

1 – a	2 – a	3 – a	4 – b	5 – b
6 – c	7 – d	8 – d	9 – d	10 – c

Coding Tasks

Coding Task 1: Shapes

Create a program that will accept numeric inputs via the command line. Based on the number of inputs, the program will decide what shape it is. All inputs in the command line should be an integer value greater than 0. For a single input, it is a circle; for two inputs, it is either a square or a rectangle; for three inputs, it is a triangle. The program will then display the shape type, details, perimeter, and area. Make sure that you follow the given UML Diagram.

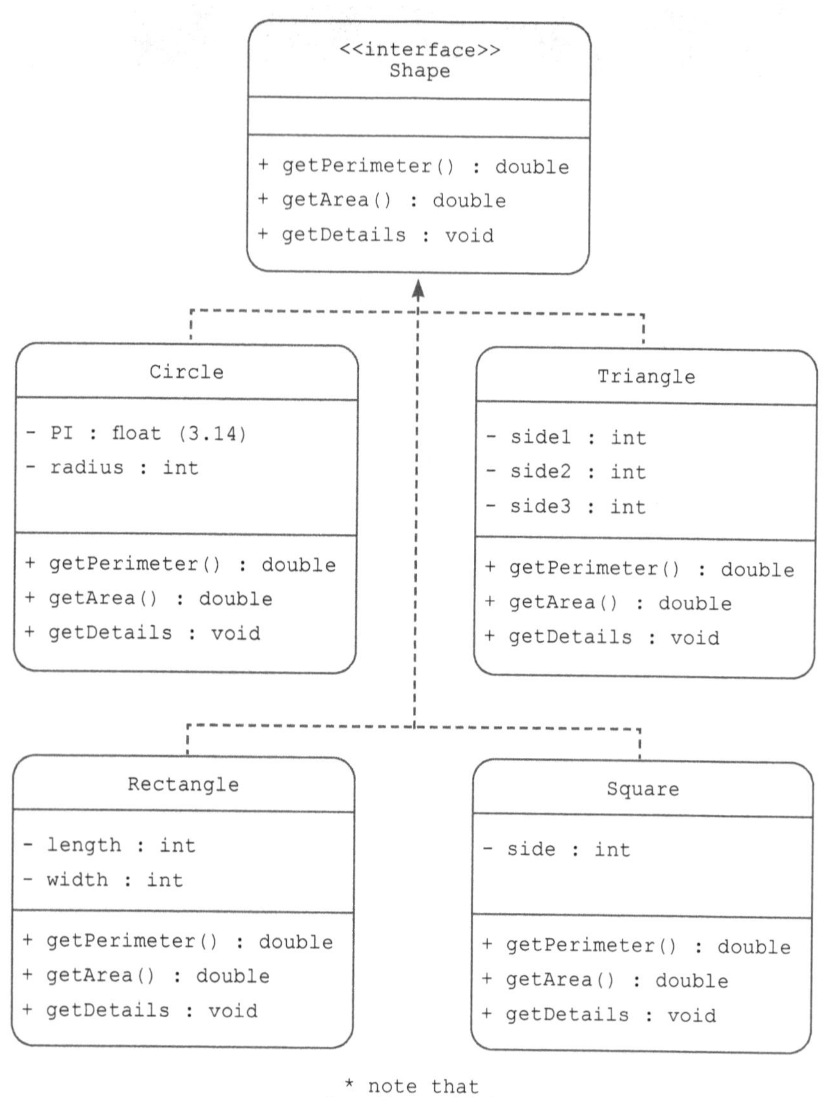

You are to create the following classes based on the given UML:

`Shapes.java`, `Circle.java`, `Rectangle.java`, `Square.java`, and `Triangle.java`. You are also required to create a test class called `TestShapes.java` that contains the `main` method, which will be compiled and run, and should accept the values via the command line.

Here are sample runtime outputs:

Sample Runtime 1:

```
java TestShapes 15 <ENTER>
Type: Circle
Radius: 15
Perimeter: 94.2
Area: 706.5
```

Sample Runtime 2:

```
java TestShapes 12 24 <ENTER>
Type: Rectangle
Length: 24
Width: 12
Perimeter: 72.0
Area 288.0
```

Sample Runtime 3:

```
java TestShapes 24 12 <ENTER>
Type: Rectangle
Length: 24
Width: 12
Perimeter: 72.0
Area 288.0
```

Sample Runtime 4:

```
java TestShapes 12 12 <ENTER>
Type: Square
Side: 12
Perimeter: 48.0
Area 144.0
```

Sample Runtime 5:

```
java TestShapes 9 15 12 <ENTER>
Type: Triangle
Side1: 9
Side2: 15
Side3: 12
Perimeter: 36.0
Area 54.0
```

Note: In computing the area of the triangle, use Heron's formula:

`https://www.mathsisfun.com/geometry/herons-formula.html`

Sample Runtime 6:

```
java TestShapes 9 15 12 8 <ENTER>
Invalid Input
```

Coding 2: Implementing an Abstract Class and Interfaces

Objective:

The objective of this task is to test students' understanding of the `abstract` keyword and `interfaces` in Java. Students will be required to design and implement a small program that uses both an `abstract` class and `interfaces` to model a simple application.

Task Description:

You are tasked with creating a small application that simulates a simple content management system (CMS) for managing different types of content: `Articles`, `Videos`, and `Podcasts`. Each content type will have certain attributes and behaviors.

Requirements:

1. **Abstract Class**: `Content`

   Create an `abstract` class named `Content` that includes:

   - Two attributes: `title` (`String`) and `author` (`String`).
   - A constructor that initializes the `title` and `author`.
   - Two abstract methods:

   `void displayInfo()` - This method should be implemented by subclasses to display the content's information.

   `Double calculateReadingTime()` - This method should be implemented by subclasses to calculate the estimated time to consume the content.

   - A concrete method:

   `void publish()` - This method should print a message indicating that the content is published.

2. **Interfaces**:

   Create two interfaces:

   - `Readable`

   Define a method `void read()` that will be implemented by content types that can be read.

- `Watchable`

  Define a method `void watch()` that will be implemented by content types that can be watched.

3. **Classes Implementing Content:**

   Class `Article`:

   - This class should extend the `Content` abstract class and implement the `Readable` interface.
   - Add an attribute `wordCount: int`.
   - Implement the `displayInfo()` method to display the `title`, `author`, and `wordCount`.
   - Implement the `calculateReadingTime()` method to calculate the reading time based on an average reading speed of 200 words per minute.
   - Implement the `read()` method to print a message indicating that the article is being read.

   Class `Video`:

   - This class should extend the `Content` abstract class and implement the `Watchable` interface.
   - Add an attribute `duration: double` representing the duration of the video in minutes.
   - Implement the `displayInfo()` method to display the `title`, `author`, and `duration`.
   - Implement the `calculateReadingTime()` method to return the duration (since the time to consume a video is just its duration).
   - Implement the `watch()` method to print a message indicating that the video is being watched.

Class `Podcast`:

- This class should extend the `Content` abstract class and implement both `Readable` and `Watchable` interfaces.
- Add attributes `duration: double` and `transcript: boolean`, indicating whether a transcript is available.
- Implement the `displayInfo()` method to display the `title`, `author`, `duration`, and whether a `transcript` is available.
- Implement the `calculateReadingTime()` method to calculate the time either based on duration (if watched) or on a word count (assume a fixed word count if the transcript is available).
- Implement both `read()` and `watch()` methods based on whether the podcast has a transcript or not.

4. **Main Method**:

- In the `main` method, create instances of `Article`, `Video`, and `Podcast`.
- Display their information using `displayInfo()`.
- Demonstrate calling the `read()` or `watch()` methods as appropriate.
- Publish each content type.

Sample Output:

```
Article: "Learn Java The Right Way", by Dr. Lawrence
Decamora

Word Count: 1200

Estimated Reading Time: 6.0 minutes

Reading Article: "Learn Java The Right Way"

Publishing content: Learn Java The Right Way

Video: "Java Tutorial", by Michelle Decamora

Duration: 30.0 minutes

Estimated Watching Time: 30.0 minutes

Watching Video: "Java Tutorial"

Publishing content: Java Tutorial

Podcast: "Java Basics", by John Doe

Duration: 20.0 minutes

Transcript Available: true

Estimated Reading Time: 5.0 minutes

Reading Podcast: "Java Basics"

Watching Podcast: "Java Basics"

Publishing content: Java Basics
```

CHAPTER 6

Exceptions and Assertions

Key Learning Objectives

- Distinguish between exceptions and errors in Java.
- Handle exceptions gracefully using `try-catch` and `finally` blocks.
- Utilize multiple catch blocks and parameterized `try` to handle different exception types.
- Apply the handle-or-declare rule and override methods while honoring exception conditions.
- Create custom `Exception` subclasses for application-specific scenarios.
- Validate assumptions and invariants during development with assertion checks.

In Java programming, building robust and reliable applications is essential for delivering a smooth and uninterrupted user experience. To achieve this, Java provides powerful mechanisms — exceptions and assertions — that help you gracefully handle unexpected situations and validate your code's assumptions. Exception handling lets you respond to problems, such as invalid input or unavailable resources, without causing your application to crash, making your code more resilient and easier to debug and maintain. Assertions, on the other hand, serve as an additional layer of assurance during development, allowing you to check conditions and invariants to help identify bugs early and avoid future issues. In this chapter, we'll explore the `Exception` and `Error` hierarchy in Java, how `try-catch` blocks, `finally`, and `throw` work, when and how to create custom exceptions, and how assertions can be effectively used to produce more reliable and robust code.

6.1 The `Exception` Class and the `Error` Class

An `Exception` is an object representation of an abnormal condition that arises in a code sequence either at compile time or at runtime. The `Exception` object handles unexpected situations like network failure, misspelled or missing file name, printers that suddenly go offline, etc. The `Exception` class is a subclass of the `Throwable` class that represents errors that can be caught and handled.

The `Error` class is a subclass of `Throwable` that represents serious problems in the Java runtime environment, which applications are not expected to handle.

Exceptions can be classified either as a checked exception that can be encountered at compile time or as an unchecked exception that can be encountered at runtime. Unchecked exceptions are commonly known as `RuntimeException`.

6.1.1 Why Do We Need to Have Exception Handling?

The following are the reasons why Exception handling is needed:

1. To notify the users, in layman's terms, of what happened during a program crash
2. To enable the users to save all, if not most, unsaved data during a program crash
3. To enable users to exit the program gracefully

Let's see what a program crash looks like if an exception is thrown and not handled. Here's a sample code called `AddArgs1.java` that will accept integer values via the command line, add them up, and display the sum.

```
// AddArgs1.java
public class AddArgs1 {

    public static void main(String[] args) {

        int sum = 0;
        for (String arg : args) {

            sum += Integer.parseInt(arg);
        }
        System.out.println("Sum: " + sum);
    }
}
```

Runtime execution via command line:

```
java AddArgs1 10 20 30 40
Sum: 100
```

The output looks good because the values passed via the command line are all integer values. Now let's try to do another runtime. This time, let's include non-integer values in the command line.

```
java AddArgs1 10 2.2 thirty 40

Exception in thread "main" java.lang.NumberFormatException:
For input string: "2.2" at java.lang.NumberFormatException.
forInputString(Unknown Source)

        at java.lang.Integer.parseInt(Unknown Source)

        at java.lang.Integer.parseInt(Unknown Source)

        at AddArgs1.main(AddArgs1.java:8)
```

When an exception is thrown and not handled, a stack trace will appear. A stack trace tells the user information about the thrown exception. The last line of the stack trace is:

```
at AddArgs1.main(AddArgs1.java:8)
```

This means that the exception originated in line 8 of the source code **AddArgs1.java,** in the main method.

6.2 Java's Approach: The Call Stack Mechanism

If a method is unable to complete the task in the usual way, Java allows it an alternative exit route. It does not seek to return a value from the method where the exception object occurred. Instead, it "throws" an object encapsulating the error information. Java then begins a search for a handler that can deal with that error.

The following code snippet shows the process of the call stack mechanism:

```
1 // SomeClass.java
2 public class SomeClass {
3
4   public static void main(String[] args) {
5
6       // unhandled exception will be thrown
7       // here from line 15, this will cause
8       // the abnormal termination of the program
9       method1();
10  }
11  public static void method1() {
12
13      // unhandled exception object from
14      // line 21 will be rethrown to line 9
15      method2();
16  }
17  public static void method2() throws Exception {
18
19      // this line will throw the
20      // Exception object to line 15
21      throw new Exception();
22  }
23 }
```

Possible output would be:

```
Exception in thread "main" java.lang.Exception
at SomeClass.method2(SomeClass.java:21)
at SomeClass.method1(SomeClass.java:15)
at SomeClass.main(SomeClass.java:9)
```

All Java SE applications always start at the `main` method. Inside the `main` method, at line 9, `method1()` was called; and inside `method1()` in line 15, `method2()` was called. However, in line 21 of `method2()`, an unhandled `Exception` was thrown. The said unhandled `Exception` object inside `method2()` will be thrown back to the calling method inside `method1()` in line 15. Since inside `method1()` the thrown `Exception` object from `method2()` is still unhandled, it will be thrown again back to the `main` method at line 9. Finally, inside the `main` method, once the `Exception` object is still unhandled, then an abnormal termination will take place, printing out the stack trace.

Figure 6.1 Image that shows the flow of the call stack mechanism

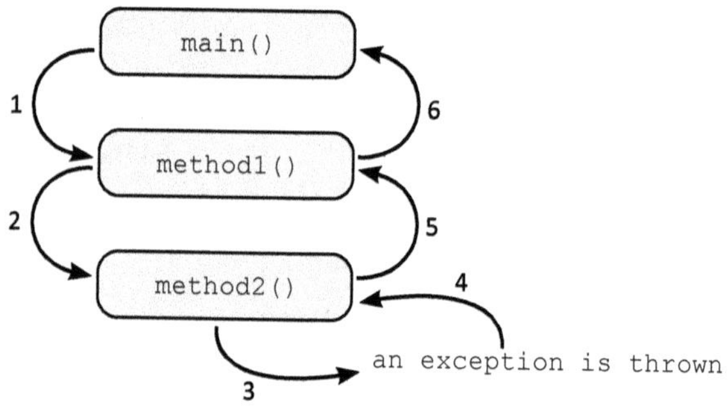

Here's a clear explanation of the Call Stack Mechanism in Java with the image in Figure 6.1 and sample code `SomeClass.java`.

The numbers in the image correspond directly to the flow of method calls and eventual Exception propagation:

Step 1 — main() is called first

- The Java Virtual Machine starts executing from `main()`.

```
public static void main(String[] args) {
    method1();
}
```

Step 2 — main() calls method1()

- The stack grows with a new stack frame for `method1()`.

```
public static void method1() {
    method2();
}
```

Step 3 — `method1()` then calls `method2()`

- The stack grows again with another stack frame for method2().

```
public static void method2() throws Exception {
    throw new Exception();
}
```

Step 4 — Inside `method2()` an `Exception` is thrown

```
throw new Exception();
```

Step 5 — `Exception` returns back to `method1()`

- `method1()` has no `try-catch` block to handle it; therefore, it exits and drops from the stack.

```
// Exception not caught here
```

Step 6 — Exception then reaches `main()`

- The `Exception` is not dealt with in `main()` either.
- So the main stack frame exits, and the program terminates abnormally.

```
// Exception not caught here either
```

Figure 6.1 illustrates this flow — starting from `main()` (1), then `method1()` (2), then `method2()` (3), then an `Exception` is thrown (4), and the stack starts unwinding upward (5 and 6).

Here's another visual flow.

```
main()
  ↓ calls
method1()
  ↓ calls
method2()
  ↓ Exception is thrown here
    Exception propagates back up:
    └─► method2() exits
    └─► method1() exits
    └─► main() exits
    └─► Unhandled Exception -> Program Terminates
```

The five keywords used for Exception handling or Exception declaration are:

- try
- catch
- finally
- throw
- throws

6.3 General Syntax of an Exception-Handling Block

The syntax below demonstrates a structured approach to handling exceptions in Java using the `try-catch-finally` block. The `try` block contains code that might throw an exception, allowing you to monitor and catch potential errors during program execution. Each `catch` block is designed to handle a specific type of exception (`ExceptionType1`, `ExceptionType2`, etc.), enabling the program to recover gracefully or provide meaningful error messages when exceptions occur. The `exOb` (exception object) parameter in each `catch` block represents the exception object, giving access to methods and details about the error. The `finally` block contains code that executes regardless of whether an exception was thrown or handled, ensuring cleanup actions, like closing resources, always occur. This structure ensures robust error management while maintaining application stability.

```
try {
   // block of code to monitor for errors or statements
   // that might cause an exception
}
catch(ExceptionType1 exOb) {
   // exception handler for ExceptionType1 and / or
   // statements to handle the Exception object
}
catch(ExceptionType2 exOb) {
   // exception handler for ExceptionType2 and / or
   // statements to handle the Exception object
}
finally {
     // block of code to be executed even if no Exception
     // occurred
}
```

6.3.1 The `try-catch` Block

The `try-catch` block is Java's way of handling the `Exception` object. Here is the revised code for adding command-line arguments:

```
// AddArgs2.java
 1 public class AddArgs2 {
 2
 3     public static void main(String[] args) {
 4
 5         try {
 6
 7             int sum = 0;
 8             for (String arg : args) {
 9
10                 sum += Integer.parseInt(arg);
11             }
12             System.out.println("Sum: " + sum);
13         }
14         catch(NumberFormatException nfe) {
15
16             System.err.println("One of the values"
17                     + " is not an integer.");
18         }
19     }
20 }
```

In lines 5 to 13, you have the `try` block, which serves as a test area. Each line of the code is run, and if an exception occurs, we move to the `catch` block. If an `Exception` occurs, the `catch` block in lines 14 to 18 will run. But if the code inside the `try` block runs smoothly, then the `catch` block will not execute.

```
java AddArgs2 10 2.2 thirty 40
One of the values is not an integer.
```

This code (**`AddArgs2.java`**) is an improved version of the previous one (**`AddArgs1.java`**). But the problem with this scenario is in the `try` block. The entire body of the `main` method was placed inside the `try` block. Placing thousands of lines of code and method calls inside a single `try` block is not recommended. It's less about runtime speed and more about writing clean, focused code that makes it easier to identify where an exception might occur. While the overhead of simply using a `try` block is minimal, performance costs become noticeable if an exception is actually thrown.

A better approach is to include only the problematic line or lines of code inside the `try` block and not include the entire body of the `main` method. Here's an improved version:

```
// AddArgs3.java
public class AddArgs3 {

    public static void main(String[] args) {

        int sum = 0;
        for (String arg : args) {

            try {

                sum += Integer.parseInt(arg);
            }
            catch(NumberFormatException nfe) {
```

```

                    System.err.println(arg + " is not"
                      + " an int and will not "
                      + " be included in the sum ");
            }
        }
        System.out.println("Sum: " + sum);
    }
}
```

```
java AddArgs3 10 2.2 thirty 40
2.2 is not an int and will not be included in the sum
thirty is not an int and will not be included in the sum
Sum: 50
```

6.3.2 The `finally` Block

The `finally` block is an optional code block that is usually placed after the last `catch` block. The `finally` block contains code that executes regardless of whether there is an exception object or not, such as closing resources like files or network connections. It ensures that important cleanup actions are always performed, maintaining resource integrity and preventing potential resource leaks. Here's a sample code snippet that demonstrates the use of the `finally` block.

```
try {

  startFaucet();
  waterLawn();
}
```

```
catch (BrokenPipeException bpe) {
  logProblem(bpe);
}
finally {
  stopFaucet();
}
```

We have a `try` block, a `catch` block, and a `finally` block. Now, the placement of the `finally` block is usually found at the last catch block. When your code snippet executes, you will enter the `try` block first and call `startFaucet()` and `waterLawn()`.

After calling the methods `startFaucet()` and `waterLawn()`, assuming there's no exception between these two method calls, your `catch` block will not run. Next, it will execute the body of the `finally` block, and it will end up calling the `stopFaucet()` method.

In an alternative scenario, this time, let's call again the methods `startFaucet()` and `waterLawn()`. If an exception object is thrown in between these two methods `startFaucet()` and `waterLawn()`, say a pipe is broken, which is represented by a `BrokenPipeException`, the `catch` block provided in the code will catch that `BrokenPipeException` using `bpe` as a reference variable, and then it will call the method `logProblem(bpe)`, which will eventually log the problem, and then finally call the method `stopFaucet()`.

What will happen to the flow of the code if we put `return` inside the catch block? Will it still call the `finally` block?

```
public class TestFinally1 {

  public static void main(String[] args) {

    System.out.println("Inside the main method..");
    doStuff();
    System.out.println("After doStuff() call..");
  }

public static void doStuff() {

  System.out.println("Inside doStuff");
  try {

    System.out.println("Inside the try block..");
    throw new Exception();
  }
  catch(Exception e) {

    System.out.println("Exception caught..");
    return;
  }
  finally {
    System.out.println("Inside the " +
                        "finally block..");
  }
 }
}
```

If you compile and run the above code, it will give you this output:

```
Inside the main method..
Inside doStuff
Inside the try block..
Exception caught..
Inside the finally block..
After doStuff() call..
```

In this sample code, take a few moments to analyze why we got this output. Remember that if the `finally` block is present, it always runs regardless of the circumstances. In this case, before the `return` statement in line 21 is called, the `finally` block gets to run.

Here's another example:

```
public class TestFinally2 {

   public static void main(String[] args) {

    System.out.println("Enter the for loop..");
    for (int i = 1; i <= 5; i++) {

       try {

        if (i == 2)
           break;
        System.out.println("i = " + i);
      }
      catch(Exception e) {

       System.out.println("inside catch block..");
      }
      finally {
```

```
19        System.out.println("inside the " +
20                           "finally block..");
21        }
22      }
23      System.out.println("after the loop..");
24    }
25  }
```

If you compile and run this code, this will give you this output:

```
Enter the for loop..
i = 1
inside the finally block..
inside the finally block..
after the loop..
```

In line 10, you have a `break` statement called inside a `for`-loop. We know that once the `break` statement is encountered inside a loop, you're supposed to end the loop. That is still true, but as you can see, we have a `finally` block. We have to execute the `finally` block regardless of the circumstances. So, before we `break` out of the for loop, we have to run the `finally` block first.

The `finally` block in Java is generally executed regardless of what happens within the `try` or `catch` blocks. However, there are a few scenarios where it might not run. One common situation is if the `System.exit(0);` method is called within the `try` or `catch` blocks, causing the JVM to terminate immediately. Another case would be if the program crashes due to a fatal error, such as running out of memory or encountering a severe system-level failure, which prevents the `finally` block from executing.

There are specific scenarios where `finally` may not execute if `System.exit()` is called:

```
try {
    System.exit(0);
} catch (Exception e) {
    // ...
} finally {
    // this will NOT execute
}
```

The JVM exits immediately and `finally` has no chance to execute.

6.4 The `Exception` Hierarchy in Java

In Java, an exception is an instance of a class derived from `Throwable`.

Figure 6.2 Java Exception Hierarchy

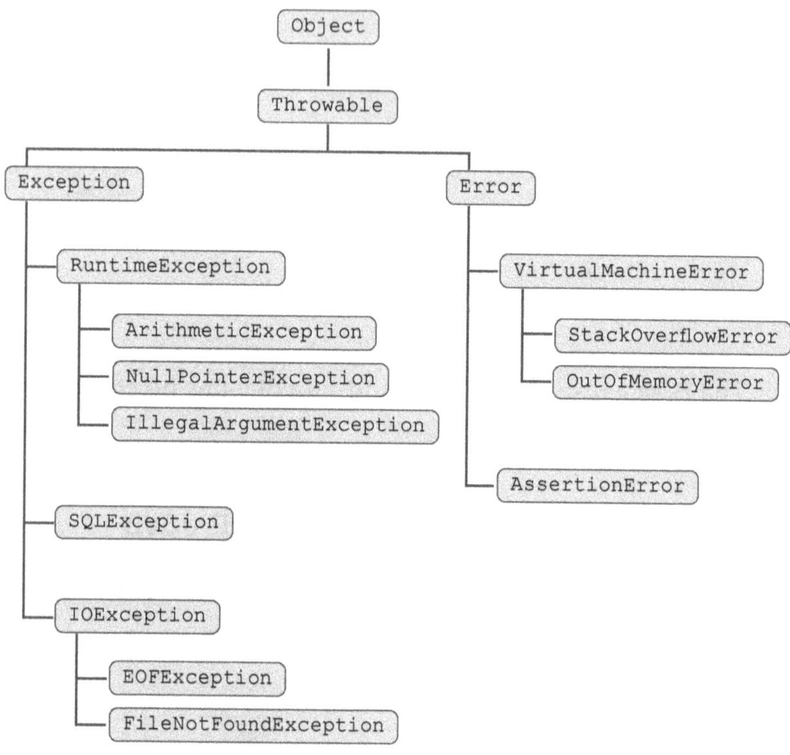

Table 6.1 provides a list of the most commonly used Exceptions and their description.

Table 6.1 Commonly used Exceptions and their description

Exception Name	Description
NullPointerException	Thrown when an application attempts to access a method or field with a null object reference.
ArrayIndexOutOfBoundsException	Thrown to indicate that an array has been accessed with an illegal index.

Exception Name	Description
`ArithmeticException`	Thrown when an illegal arithmetic operation occurs, such as division by zero.
`ClassCastException`	Thrown when an object is cast to a subclass that it is not an instance of.
`IOException`	Thrown when an I/O operation fails or is interrupted, such as file reading or writing errors.
`FileNotFoundException`	A subclass of IOException, thrown when an attempt to open a file that does not exist fails.
`NumberFormatException`	Thrown when attempting to convert a string to a numeric type fails due to an invalid format.
`IllegalArgumentException`	Thrown when a method receives an invalid argument.
`InterruptedException`	Thrown when a thread is interrupted while it is sleeping, waiting, or otherwise paused.
`SQLException`	Thrown to indicate errors related to database access or other SQL-related issues.

6.5 Multiple Exceptions in a `catch` Block

Prior to Java SE 7, if you needed to catch several `Exception` objects of different types and do the same error management, you would do it this way:

```
try {
 // some codes that throws some Exception object
}
catch(IOException ex) {
 logProblem(ex);
}
catch(SQLException ex) {
 logProblem(ex);
}
```

In Java SE 7, multiple `Exception` objects can be caught inside a single `catch` block like this:

```
try {
 // some codes that throws some Exception object
}
catch(IOException | SQLException ex) {
 logProblem(ex);
}
```

In this sample code, if the `try` block throws an `IOException` or a `SQL Exception` object, either `Exception` will be caught by the same `catch` block. This code is shorter, cleaner, and less redundant compared to the first one.

6.6 The Parameterized `try` Block (Try-With-Resources)

Starting Java SE7, a `try` block can now have resource parameters. These resources are objects that need to be closed after execution. Any object that implements the interface `java.lang.AutoCloseable`, which includes all types of objects that implement the interface `java.io.Closeable`, can be used as a resource that can be passed as a parameter to a `try` block.

Here is a code from before Java SE 7 that does not implement a parameterized `try` block:

```
import java.io.*;
public class SampleIO {

 public static void main(String[] args) {

     try {

         BufferedReader in =
                 new BufferedReader(new
                 InputStreamReader(System.in));
         System.out.println("What is your name?");
         String name = in.readLine();
         System.out.println("How old are you?");
         int age = Integer.parseInt(in.readLine());
         System.out.println("Hello " + name +
                 ", next year you will be " + (age + 1)
                 + " years old.");
         in.close();
     }
     catch(IOException ioe) {

         System.err.println("An exception occurred.");
     }
 }
}
```

In line 8, we see that the `BufferedReader in` object was created and used in the rest of the code, but at the end of the `try` block in line 18, the `in` object was closed.

Here is a code that includes Java SE 7's try-with-resources.

```
1 import java.io.*;
2 public class SampleIO2 {
3
4   public static void main(String[] args)
5                                     throws IOException {
6
7       try(BufferedReader in = new BufferedReader(new
8                   InputStreamReader(System.in))) {
9
10          System.out.println("What is your name?");
11          String name = in.readLine();
12          System.out.println("How old are you?");
13          int age = Integer.parseInt(in.readLine());
14          System.out.println("Hello " + name +
15                      ", next year you will be "
16                      + (age + 1) + " years old.");
17      }
18  }
19 }
```

Notice that in lines 7 and 8, you have a `try` block that has parameters in it. The object `BufferedReader in` that was declared inside the `try` block parameter is considered a resource and will be closed automatically after the `try` block is finished in line 17. No need to call the `in.close()` method because all resources declared inside the parameterized try block will be closed as long as those objects are `AutoCloseable` objects. This will lessen the risk of having memory leaks.

6.7 The Handle or Declare Rule

There are two approaches in dealing with an `Exception` object: you can either handle it or you can declare it. Handling the `Exception` object means you can use the `try-catch-finally` blocks like we have done so far.

Declaring an `Exception` object, on the other hand, offers an entirely new approach by using the keyword `throws`. Sample methods for declaring an `Exception` object:

```
void doStuff() throws IOException {}
void someMethod() throws SQLException, IOException {}
```

Here is a sample code that declares an `Exception` object.

```
// SampleIO3.java
import java.io.*;
public class SampleIO3 {
  public static void main(String[] args)
                                  throws IOException {
      BufferedReader in =
              new BufferedReader(new
              InputStreamReader(System.in));
      System.out.println("What is your name?");
      String name = in.readLine();
      System.out.println("How old are you?");
      int age = Integer.parseInt(in.readLine());
```

```
15         System.out.println("Hello " + name +
16               ", next year you will be " + (age + 1)
17               + " years old.");
18         in.close();
19
20     }
21 }
```

Declaring an Exception is only done for checked exceptions and uses the `throws` keyword after the method signature, just like in line 6. It is like issuing a waiver to allow the checked exception to pass the compilation process. `RuntimeException` objects or unchecked exceptions, on the other hand, should not be declared; instead, `RuntimeException` objects can be handled if desired. If an exception is declared and actually occurs at runtime, an abnormal termination will still happen. In most cases, the declaration of `Exception` is only done during the software development stage and is slowly converted to handled exceptions during software quality assurance and testing.

6.8 Improved Final or Effectively Final Variable Rethrow (JDK 11+)

Starting with JDK 11, the compiler allows more flexible use of rethrowing exceptions when the caught exception is effectively `final`. This allows rethrowing without explicit casting, improving readability. For example:

```
try {
    // risky operation
} catch (IOException | SQLException e) {
    throw e; // no need to cast explicitly
}
```

In earlier versions, developers needed to cast e to a common superclass, which cluttered code and introduced unnecessary verbosity.

Here are two examples of the before and after improvement of *final variable rethrow* feature.

Before JDK 11:

```
public class BeforeJDK11 {
    public static void main(String[] args)
                     throws Exception {
        try {
            processData(null);
        } catch (Exception e) {
            // Need a new variable to rethrow
            Exception ex = new Exception("Updated" +
                     " message: " + e.getMessage());
            throw ex;
        }
    }

    static void processData(String input) throws Exception {
      if (input == null) {
        throw new Exception("Input cannot be null!");
      }
    }
}
```

After JDK 11:

```
public class AfterJDK11 {
  public static void main(String[] args) throws Exception {
    try {
        processData(null);
```

```
        } catch (Exception e) {
            // Reassign and rethrow the same variable
            e = new Exception("Updated message: " +
            e.getMessage());
            throw e; // Allowed in JDK 11+
        }
    }

    static void processData(String input)
                            throws Exception {
        if (input == null) {
            throw new Exception("Input cannot be null!");
        }
    }
}
```

Before JDK 11, you could not reassign and then rethrow the same variable without using a separate variable or marking it as `final`. This improvement makes exception handling more flexible and code easier to maintain.

In JDK 11+, you no longer need to create a second variable just to rethrow an updated exception. This makes your code shorter and easier to read.

6.9 Rules on Overriding Methods and Exceptions

Consider Figure 6.3.

Figure 6.3 Overriding Methods with Exceptions

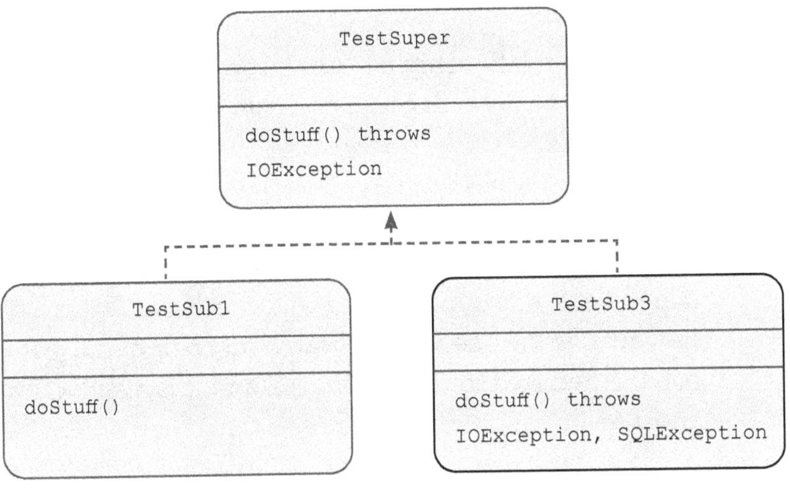

TestSub2

doStuff() throws
FileNotFoundException,
EOFException

Let's break this down into three cases.

- **Case 1:** The relationship of `TestSuper` and `TestSub1`: In this inheritance relationship, everything is well. It will compile and run because the superclass `TestSuper` declared an `IOException`, and the subclass `TestSub1` has no declared exceptions. This is acceptable.
- **Case 2:** The relationship between `TestSuper` and `TestSub2`: This too is acceptable. In the superclass `TestSuper`, there is the `IOException` declared, while the subclass `TestSub2` declares two exceptions, the `FileNotFoundException`, and

the EOFException. This is acceptable because both FileNotFoundException and the EOFException are subclasses of IOException.

- **Case 3:** The relationship of TestSuper and TestSub3: This relationship will cause an error. In the superclass TestSuper, there is the IOException declared, while the subclass TestSub3 declares two exceptions, IOException and SQLException. IOException is allowed because it was already declared in the superclass TestSuper. The SQLException, however, is the one that will cause an error because it is NOT declared in the superclass, and it is also NOT a subclass of IOException.

6.10 Create Your Own Exception

To create your own Exception object, you simply create a class that extends the class Exception. Since we have both checked and unchecked (RuntimeException), let us see how we can create our own checked and unchecked exceptions.

Creating your own checked Exception:

```
class BadTastingFoodException extends Exception {
    // put code here
}
```

With this example, you have just created a new Exception class called BadTastingFoodException. Now, let's put it to use.

```
// MyOwnException.java
class BadTastingFoodException extends Exception  {
  // you can leave this part empty or call the constructor
    // of the Exception class to pass a String message.
}

public class MyOwnException {

  private static boolean candy = false;
  private static boolean sourBall = true;

  public static void main(String[] args) {
      try {
          System.out.println("Inside try block of "
                                     + " candy.");
          isSour(candy);
      }
      catch(BadTastingFoodException btfe) {
          System.out.println("Inside catch block "
                                     + "of candy.");
      }
      try {
          System.out.println("Inside try block "
                                     + " of sourBall.");
          isSour(sourBall);
      }
      catch(BadTastingFoodException btfe) {
          System.out.println("Inside catch block "
                                     + " of sourBall.");
      }
  }
  private static void isSour(boolean taste)
```

```
        throws BadTastingFoodException  {
    if (taste)
        throw new BadTastingFoodException();
    }
}
```

Sample Output:

```
java MyOwnException
Inside try block of candy.
Inside try block of sourBall.
Inside catch block of sourBall.
```

Checked exceptions are checked at compile time. So when you create your class and that class extends the class Exception, you're creating your own checked exceptions, which means this exception can be seen during compile time.

6.11 Creating Your Own Unchecked `RuntimeException`

As for the creation of your own `RuntimeException`, you have to extend the `RuntimeException` class. Here's an example:

```
// MyOtherExceptionExample.java
import java.util.Scanner;

class LoginException
 extends RuntimeException { // unchecked exception
}

```

```
public class MyOtherExceptionExample {
  private static String username = "admin";
  private static String password = "password";

  public static void main(String[] args) {
    // try-with-resources ensures Scanner
    // is closed automatically
    try (Scanner s = new Scanner(System.in)) {
      System.out.println("Enter your username: ");
      String uname = s.nextLine();

      System.out.println("Enter your password: ");
      String pword = s.nextLine();

      if (uname.equals(username)
        && pword.equals(password)){
        System.out.println("Access Granted");
      } else {
        throw new LoginException();
      }
    } catch (LoginException ae) {
      System.err.println("Incorrect username "
                + " / password");
    }
  }
}
```

As you can see, the class LoginException is considered an unchecked exception because it extends the RuntimeException class. This means the compiler cannot see this type of exception during compile time. This also means that the best way to deal with these types of

`Exception` classes is by using the `try-catch`-block and not just using the keyword `throws`.

6.12 Assertions

Assertions check the programmer's assumptions. Assertions in Java are a crucial tool for validating a programmer's assumptions about how code should behave at specific points during execution. By embedding assertion checks directly into the code, developers can proactively identify and address bugs early in the development process, ensuring that program logic aligns with expected conditions. Assertion checks can be programmatically placed in code, but whether they are executed depends on the runtime configuration. Assertions can run in two modes: enabled or disabled. When enabled (using the `-ea` or `-enableassertions` flag), the assertion condition is evaluated, and if it fails, an `AssertionError` is thrown. When disabled (the default), assertions are ignored by the JVM and have no effect. This flexibility allows assertions to be used for debugging during development without impacting performance in production, making them a powerful approach for building more reliable and error-free code.

6.12.1 Syntax for the Use of the `assert` Keyword

```
assert <boolean_expression> ;
assert <boolean_expression> : <detail_expression> ;
```

If you want to customize the displayed message, you can customize the ***`<detail_expression>`*** part. If the

<boolean_expression> returns a false value, it will throw an `AssertionError`.

Here's a sample code:

```
// TestAssert.java
public class TestAssert  {
  public static void main(String[] args)  {
      int hour = Integer.parseInt(args[0]);
      int mins = Integer.parseInt(args[1]);

      assert hour > 0 && hour < 24 :
                  "Invalid hour value";
      assert mins > 0 && mins < 60 :
                  "Invalid mins value";

      System.out.println(hour + ":" + mins);
  }
}
```

Sample Output:

```
java TestAssert 15 30
15:30
```

Here's another one:

```
java TestAssert 15 300
15:300
```

Notice the inconsistency? The second argument (300) should have caused an error. Why is there no error at runtime?

Java version 1.4 has a different behavior compared to Java 1.5 and onwards as shown in Table 6.2.

Table 6.2 Default assertion setting in Java 1.4 and Java 1.5

	Compile Time	Runtime
java 1.4	javac -source 1.4 Filename.java	Default behavior: **enabled** to disable: java **-da** Filename
java 1.5 and 1.6	java Filename. java	Default behavior: **disabled** to enable: java **-ea** Filename

In the last runtime example, there is no error because the runtime environment used was Java 1.5 or above. This means that during program execution or runtime, the assertion checking is disabled in your code, thus ignoring the assertion check. If you want to enable the assertion checks, you may want to rerun your application this way:

```
java -ea TestAssert 15 300

Exception in thread "main" java.lang.AssertionError:
Invalid mins value

at TestAssert.main(TestAssert.java:9)
```

By putting **`-ea`** or **`-enableassertions`**, as a VM (Virtual Machine) option, you're actually enabling the assertion check on your application at runtime.

Also, you need to note that the error message states that it is an `AssertionError` and not an `Exception`, therefore, you need to recheck or revalidate either your condition or your assumptions, or your input values.

6.12.2 Assertion Support and Unicode Output (JDK 13+)

From JDK 13 onwards, assertions can also display Unicode characters properly in their error messages, which means you can include symbols, emojis, or text in other languages. Although the `assert` keyword remains unchanged, the overall experience when using assertions has improved due to better Unicode handling in console output and tooling support across IDEs. This makes debugging non-English or symbol-rich messages more reliable during development.

```
public class UnicodeAssertionExample {
  public static void main(String[] args) {
    int score = 45;

    // This assertion uses Unicode symbols
    assert score >= 50 :
        "✗ Score too low! 최소 점수는 50 입니다.";
    // ("최소 점수는 50입니다"
    // means "Minimum score is 50" in Korean)

    System.out.println("✓ Test passed!");
  }
}
```

Here's how you should run this code:

```
java -ea UnicodeAssertionExample
```

You can now add **Unicode** characters (like ✓, ✗, or text in other languages) to make assertion messages clearer and more expressive. This helps make debugging output more readable, especially in multilingual applications.

Before JDK 13 (Possible Output):

```
assert false : "✗ 점수가 너무 낮습니다!";
```

May appear as:

```
java.lang.AssertionError: ??? ????? ??
(depending on terminal or environment support)
```

JDK 13 and Later (Correct Output):

```
assert false : "✗ 점수가 너무 낮습니다!";
java.lang.AssertionError: ✗ 점수가 너무 낮습니다!
```

Before vs. After JDK 13 – Unicode in Assertions

Table 6.3 Comparison table of Unicode in Assertions

Before JDK 13	JDK 13 and Later
Assertion messages with certain Unicode symbols or non-Latin characters **might** display incorrectly or as garbled text in some environments.	Assertion messages correctly display Unicode symbols, emojis, and text in various languages.
✗ Limited readability in multilingual or symbol-rich error messages.	✓ Clear, expressive, and more user-friendly assertion messages.

Chapter Summary

- Learning how to handle exceptions in Java is crucial because it enables students to write robust and error-resistant programs, ensuring that the application can gracefully manage unexpected situations. Proper exception handling also improves code readability and maintenance by clearly separating error-handling logic from regular code execution.
- An `Exception` is a condition that a program can anticipate and handle, allowing the program to recover or take corrective action, while an `Error` represents a serious problem that typically indicates a failure in the runtime environment, such as `OutOfMemoryError`, and is usually not meant to be caught or handled by the application. `Exceptions` are intended for conditions that a program should catch, whereas `Errors` are usually beyond the program's control.
- This is why in Java, we use the `try-catch` block to allow the program to test a block of code that might throw an exception (the `try` block) and to handle the exception gracefully if one occurs (`catch` block), preventing the program from crashing. This structure helps isolate error-handling code and maintain the normal flow of the program even when an error arises.
- Another keyword that a Java programmer should know in exception handling is the `finally` block. It is used to execute code that must run regardless of whether an exception was thrown or caught in the associated `try` block, such as closing resources like files or network connections. It ensures that important cleanup actions are always performed, maintaining resource integrity and preventing potential resource leaks.

- The handle and declare rule in Java exception handling requires that any checked exception must either be caught and handled using a `try-catch` block (handle) or be declared in the method's signature using the `throws` keyword (declare). This rule enforces explicit error handling, ensuring that exceptions are either dealt with directly or acknowledged and passed up the call stack for higher-level methods to manage. It helps in building more reliable and predictable software by clearly outlining how exceptions should be treated.
- To create your own exception in Java, you need to write a new class that extends the `Exception` class (for a checked exception) or the `RuntimeException` class (for an unchecked exception). This custom exception class can include additional constructors and methods to provide specific error details or behavior tailored to your application's needs, or it can be just an empty class.
- When overriding a method, the overriding method cannot declare to `throw` more or broader checked exceptions than the method it overrides. However, it can declare fewer or narrower checked exceptions or any number of unchecked exceptions, ensuring that the subclass remains consistent with the exception handling expectations of the superclass.
- The `assert` keyword in Java is used to create an assertion, which is a statement that tests an assumption in the code, typically used for debugging purposes. If the assertion is false, an `AssertionError` is thrown, allowing developers to detect and correct logic errors during development.

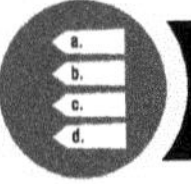

Multiple Choice Questions

1. **The following situations can cause an `Exception`, EXCEPT ________.**
   a. The JVM ran out of memory.
   b. Sending a file to a server, but the connection was interrupted.
   c. Trying to divide an integer value by an integer 0.
   d. Trying to access a misspelled file.

2. **How do you create your own `Exception`?**
   a. Create a `throws new Exception()` constructor call
   b. Create a class and extend the `Exception` class
   c. throw an `Exception` object
   d. throw an `IOException` object

3. **When you create a user-defined exception, you create a class that extends the class `Exception` from the `java.lang` package. What type of `Exception` class are you creating?**
   a. You are creating a `RuntimeException`.
   b. You are creating an unchecked exception.
   c. The exception type varies, depending on what operating system you are using.
   d. You are creating a checked exception.

4. **Fill in the blank: If your program crashes, Java will print the ___________ to your terminal, or on your output console, or on your command.com window (cmd window), which tells you the list of the methods executing when the error occurred.**

   a. compilation error messages
   b. compilation logs
   c. stack trace
   d. VM logs

5. **Given the code:**

```
class TestAssert {
  public static void main(String [] args) {
   assert(false): "more info ";
   System.out.println("after assert ");
  }
}
```

Which of the following statements is TRUE?

   a. The command-line invocation java TestAssert will produce the output more info
   b. The command-line invocation java TestAssert will produce the output after assert
   c. The command-line invocation java TestAssert will produce the output more info after assert
   d. The command-line invocation java TestAssert will produce the output after the assert more info

6. **Which of the following keywords IS NOT used for exception handling?**
   a. `try`
   b. `break`
   c. `catch`
   d. `finally`

7. **Which keyword is used to declare an `Exception`?**
   a. `try`
   b. `throw`
   c. `throws`
   d. `catch`

8. **The ________ keyword is used to manually cause an exception object to be thrown.**
   a. `try`
   b. `catch`
   c. `throw`
   d. `throws`

9. **In creating your own exception, you need to create a class that extends the class ________. This user-defined exception is considered an unchecked exception.**
   a. `Object`
   b. `Throwable`
   c. `Exception`
   d. `RuntimeException`

10. **Which code snippet below will throw an AssertionError when executed with assertion-enabled turned on?**

    a. `assert(false);`

    b. `assert(true);`

    c. `assert(false || true);`

    d. `assert(assert);`

Multiple Choice Answer Key

1 – a	2 – b	3 – d	4 – c	5 – b
6 – b	7 – c	8 – c	9 – d	10 – a

Coding Tasks

Coding Task 1:

You are to revisit the coding task #1 in Chapter 5: Shapes, and consider adding exception handling for the following scenarios:

- If any of the arguments contains a negative value, throw an `IllegalArgumentException`.
- If the argument list exceeds three values or has no value at all, throw an `IllegalArgumentException`.
- If any of the input values contain a `String`, throw a `NumberFormatException`.

Here are the new sample runtime outputs:

Sample Output #1:

```
java TestShapes -15 <ENTER>
A negative value was passed.
An IllegalArgumentException is caught.
```

Sample Output #2:

```
java TestShapes <ENTER>
Pass one to three values via command line.
An IllegalArgumentException is caught.
```

Sample Output #3:

```
java TestShapes 10 20 30 40 <ENTER>
Limit your arguments to a maximum of 3 values.
An IllegalArgumentException is caught.
```

Sample Output #4:

```
java TestShapes 10 test 20 <ENTER>
An argument contains a String value.
A NumberFormatException is caught.
```

Coding Task 2:

The Pythagorean Theorem is one of the most well-known mathematical principles. It asserts that in a right triangle, the sum of the squares of the two shorter sides is equal to the square of the longest side, known as the hypotenuse. This can be expressed as: $a^2 + b^2 = c^2$, where `a` and `b` are the lengths of the two shorter sides, and `c` is the length of the hypotenuse.

Your task is to complete the provided Java code that allows the user to input two numeric values, `double a` and `double b`. The program should then calculate and display the length of the hypotenuse `double c`. For example, if the user inputs `a = 3` and `b = 4`, the program should output `c = 5.0`.

Incorporate exception handling into your code:

- If the user inputs a negative value, throw an `IllegalArgumentException`.
- If the user inputs a non-numeric value, throw a `NumberFormatException` or an `InputMismatchException`.
- Ensure that these exceptions are caught, and display a clear and informative message to the user explaining the nature of the incorrect input.

Additionally, add comments after each line of code to describe its purpose and functionality.

CHAPTER 7

Building Java Text-Based Applications

Key Learning Objectives

- Understand the fundamentals of Java I/O streams
- Utilize `java.util.Scanner` and `String.format()` for input and output
- Perform file operations using the `File` class and file streams
- Convert string data into numeric types safely and efficiently
- Chain streams together to process data in flexible and reusable ways
- Develop text-based applications that read from and write to files

This chapter focuses on *how Java handles input and output of text-based applications,* a fundamental aspect of developing interactive and data-centric programs. Understanding *input and output (I/O) streams* is essential for managing the flow of data within your application.

Here, you'll learn about the fundamentals of streams, including key concepts such as stream classes, how data is processed within streams, and node streams for direct data handling.

The chapter also covers how *I/O stream chaining* can enhance functionality and flexibility in your code, and it delves into Java's versatile I/O capabilities. It shows you how to perform common tasks such as reading input from the keyboard, converting strings to numeric data, and using `java.util.Scanner` to accept formatted input. Additionally, you'll learn how to produce nicely formatted output with `String.format()` for a more polished appearance.

This chapter also explores *working with files* through the `File` class, explaining the nature of a file object, its connection to physical files on your hard drive, and useful methods for file testing and operations. Finally, you'll gain hands-on experience reading from and writing to files, allowing you to efficiently handle persistent data in your Java applications. By the end of this chapter, you'll have a solid understanding of how to perform a range of I/O operations to develop interactive, file-centric, and flexible Java programs.

7.1 I/O Stream Fundamentals

A *stream* is defined as a flow of data from a source to a destination or a sink. A *source* stream initiates the flow of data, while the destination (sink) terminates the flow of data. Sources are also called input streams, and sinks are also called output streams. Sources and sinks are both *node streams,* also called endpoints. These streams can be opened and closed before and after the input and output of data is finished.

There are different types of nodes such as file nodes, memory nodes, pipes, or inter-thread (pipe) nodes.

7.2 Fundamental Stream Classes

Java maintains two groups of streams, byte streams and character streams, both of which have their respective input and output stream classes.

Table 7.1 Fundamental Stream Classes - Byte Streams and Character Streams

Stream	Byte Stream	Character Stream
Source Streams	`InputStream`	`Reader`
Sink Streams	`OutputStream`	`Writer`

7.3 Data Within Streams

There are classes under the `java.io` package that allow the conversion of byte streams to Unicode text or character streams and vice versa. `Reader` and `Writer` classes handle the Unicode text, while byte streams handle raw byte data.

For each kind of *node*, whether it's a *file*, a *memory array*, or even a *network connection*, there is a specific Java class designed to handle its input and output streams.

This means that when you want to read from or write to a particular data source — like a file on your computer's hard drive or a block of memory — Java provides a specialized class to do just that.

Some classes work directly with byte streams, which deal with raw binary data, while others work with *character streams*, which handle Unicode text.

For example, `FileInputStream` and `FileOutputStream` are designed to read and write bytes from a file, whereas `FileReader` and `FileWriter` are tailored for reading and writing character data.

This lets you choose the most appropriate tool for your specific needs and makes it easier to perform I/O operations in a way that is efficient, convenient, and reliable. Table 7.2 lists common node stream classes.

Table 7.2 shows sample node streams together with their corresponding classes.

Table 7.2 Sample Node Streams - Byte Streams and Character Streams

Type	Byte Streams	Character Streams
File	`FileInputStream` `FileOutputStream`	`FileReader` `FileWriter`
Memory Array	`ByteArrayInputStream` `ByteArrayOutputStream`	`CharArrayReader` `CharArrayWriter`
Memory String	`N/A`	`StringReader` `StringWriter`
Pipe	`PipedInputStream` `PipedOutputStream`	`PipedReader` `PipedWriter`

This table provides some of Java's Node Streams, which are used for input and output operations. Node streams directly read from or write to various data sources like *files, memory arrays (this is means a block of memory that temporarily stores data while your program is running)*, or *pipes (is a way for two parts of a Java program to communicate directly with each other — it's like a tube that lets data flow from one side – the output, to the other side – the input).*

Most methods from the byte stream classes are also present in the character stream classes. That is why, in most cases, whatever functionality a byte stream can do can also be done by the character streams. The only difference between these two streams is the type of character they process. Byte streams can process non-text-based or compressed data, while character streams can process text-based or uncompressed data. Sample file types that can be processed by byte streams are JPG, PDF, MP3, AVI, etc.. For character streams, we have txt, html, css, xml, js, csv, and other uncompiled source codes like py, cpp, and java files.

7.4 I/O Stream Chaining

Figure 7.1 shows that it is possible to convert and pass data from a byte stream to a character stream and vice versa so that the JVM processes the said data in a Unicode format, then it can be returned to become non-Unicode data for either storage, display, streaming or further manipulation.

Figure 7.1 I/O Stream Chaining

Input Stream Chain

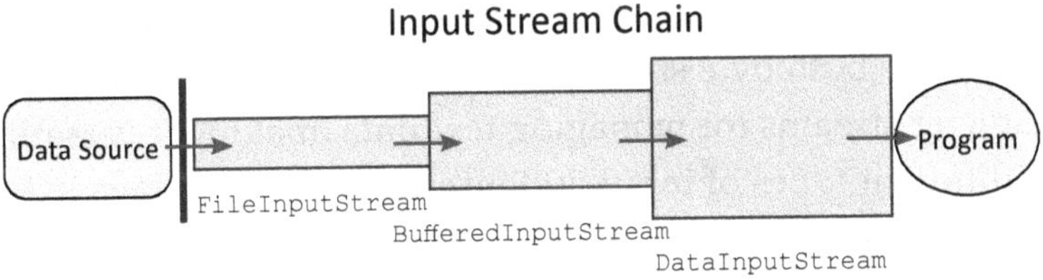

Output Stream Chain

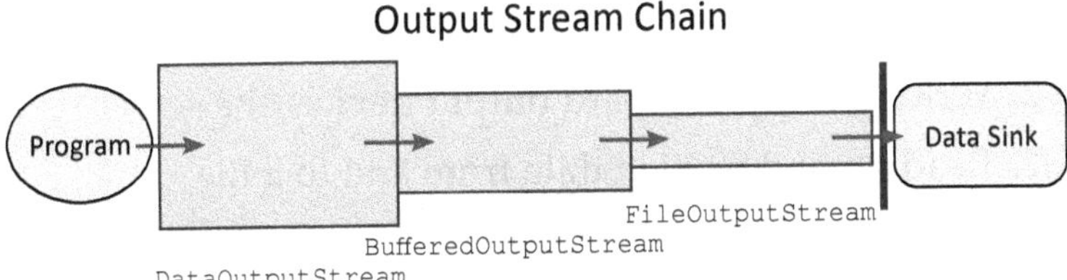

An example of this scenario is when you accept values from your keyboard. The character encoding used in Java 1.0 was ASCII (American Standard Code for Information Interchange). It encodes 128 characters using 7 bits, including English letters, digits, and some control characters. Usually, all these characters are found on your standard keyboard. In 1997, when Java 1.1 was released, one of the major changes was the character encoding. From ASCII, it was changed to Unicode to address the need for internationalization support, allowing other characters aside from the English characters to be included. Unicode is a comprehensive standard that can encode over 1.1 million characters using different encoding forms (UTF-8, UTF-16, UTF-32). It covers characters from various languages, symbols, and emojis, providing broader global support compared to ASCII. So when we accept inputs from our keyboard, which is in ASCII format, it will be piped or converted into Unicode so that the JVM can process it.

7.5 Java I/O

Java I/O (Input and Output) provides a set of classes and methods for reading data from and writing data to various data sources such as files, memory, and network connections. It supports both byte streams for handling binary data and character streams for managing text data, making it versatile for different types of input and output operations. Java I/O provides a way to do the following:

1. Reading input from the standard input device (the keyboard)
2. Writing to the standard output device (the screen)
3. Reading and writing data from and to a file.

7.6 Reading Input via Keyboard

Reading from the standard input device uses a predefined object called `System.in.` The `System.in` is an input stream object that is, by default, open and ready to supply input data. Typically, this stream supports data input from the keyboard. We use this code snippet to set up our input stream:

```
BufferedReader br = new
        BufferedReader(new InputStreamReader(System.in));
```

The `System.in` object accepts an ASCII input from your keyboard, then it passes the input stream object to the `InputStreamReader`. The `InputStreamReader` is a bridge from byte streams to character streams. It reads bytes and decodes them into Unicode characters.

The `BufferedReader` class, on the other hand, reads text from a character-input stream, buffering characters so as to provide for the efficient reading of characters, arrays, and lines. You can then use the method `readLine()` under `BufferedReader` to read a line of input from the standard input device.

Sample Program:

```
import java.io.*;
public class KeyboardInput {
  public static void main(String[] args) throws IOException {
        BufferedReader br = new BufferedReader(new
                  InputStreamReader(System.in));
```

```
        System.out.println("Enter text " +
                "and press Ctrl + D or Ctrl + Z to end.");
        String str = br.readLine();
        while(str != null) {

            System.out.println("Read: " + str);
            str = br.readLine();
        }
        br.close();
    }
}
```

In the Windows OS command prompt, `Ctrl + Z` is equivalent to the `null` character, while in the Unix / Linux / Mac OS terminal window, that would be equivalent to `Ctrl + D`. In the given code, the `null` character is needed to terminate the loop.

7.6.1 The Conversion of `String` to Primitive Data Types

It is common to encounter scenarios where a value stored as a string needs to be converted into a primitive data type for calculations or logic processing. Converting strings to numeric and `boolean` types is essential when working with data in Java. Here are different ways to convert a `String` value into different numeric and `boolean` types.

1. Conversion of `String` to integral values

To convert a `String` representing a whole number into integral data types like `byte`, `short`, `int`, or `long`, the respective parse methods of the wrapper classes can be used. Consider the following code snippets:

```
String strNum = "100";
byte b = Byte.parseByte(strNum);    //Converts to byte
short s = Short.parseShort(strNum);//Converts to short
int i = Integer.parseInt(strNum);   //Converts to int
long lNum = Long.parseLong(strNum); // Converts to long
```

These methods are particularly useful when working with data input as strings, such as user inputs or file data, that need to be processed numerically.

2. Conversion of `String` to floating-point values

Strings representing decimal numbers can be converted to `float` or `double` using the corresponding methods. Examine the code snippets below:

```
String sNum = "1.1";
float f = Float.parseFloat(sNum);   //Converts to float
double d = Double.parseDouble(sNum); //Converts to double
```

These conversions allow precise handling of numbers with fractional values, such as when performing calculations in scientific or financial applications.

3. Conversion of `String` to boolean values

When dealing with logical data stored as strings, the `Boolean.parseBoolean()` method converts the string into a `boolean` value. The method returns `true` if the string equals "`true`" (case-insensitive) and `false` for any other value:

```
String str = "true";
boolean b = Boolean.parseBoolean(str); // Converts to boolean
```

This is useful for processing flags or toggles represented as strings, such as configuration settings or user preferences.

By understanding these conversions, developers can seamlessly transform string-based inputs into appropriate data types, enabling accurate and efficient computations.

7.6.2 Accepting Formatted Input By Using the `java.util.Scanner` Class – Revisiting the `Scanner` Class

As of Java version 5.0, accepting formatted input is now possible by using the `Scanner` class under the `java.util` package. The `Scanner` class can be used to accept formatted inputs from any data source using your `Scanner` object.

Here is a sample code that uses the `Scanner` class to accept input. The filename is `ScannerExample.java`.

```
import java.util.Scanner;

public class ScannerExample {

public static void main(String[] args) {

  Scanner s = new Scanner(System.in);
  System.out.println("What is your name?");
  String name = s.nextLine();

  System.out.println("How old are you?");
  int age = s.nextInt();

  System.out.println("Enter your GPA: ");
  double gpa = s.nextDouble();

  System.out.println("Hello " + name
```

```
      + ", next year you'll be "
      + (age + 1) + " years old." + "\nGPA is: " + gpa);
      s.close();
    }
  }
```

The runtime output would be:

```
What is your name?
Lawrence <ENTER>
How old are you?
21 <ENTER>
Enter your GPA:
1.75 <ENTER>
Hello Lawrence, next year you'll be 22 years old.
GPA is: 1.75
```

The Scanner class has methods that will allow formatted inputs as shown in Figure 7.2.

Figure 7.2 Scanner methods

```
Data Type   The Scanner nextXxx() methods
===========================================
byte        nextByte()
short       nextShort()
int         nextInt()
long        nextLong()
float       nextFloat()
double      nextDouble()
boolean     nextBoolean()
char        next().charAt(0) -- yes, we don't have nextChar()
String      next() or nextLine()
```

7.6.3 Displaying Formatted Output Using `String.format()` Method

In the good old days of Java 1.4, it would take several lines of code to format an output the way you wanted it to appear. However, since the birth of Java 5.0, formatting output has never been easier. With the use of `String.format()` or the `System.out.printf()` methods, you can use these methods to format an output.

Common formatting codes are listed in Table 7.3.

Table 7.3 String Format Specifiers

Code	Description
%s	Formats the argument as a String, usually by calling the toString() method on the object.
%d %o %x	Formats an integer as a decimal, octal or hexadecimal value
%f %g	Formats a floating point value, %g uses scientific notation
%n	Inserts the newline character to the String or stream.
%%	Inserts the % character to the String or stream.

Here's an example:

```
import java.io.*;
import java.util.Scanner;

public class DisplayAverage  {
 public static void main(String[] args) throws IOException {
     Scanner s = new Scanner(System.in);
     System.out.println("Enter first number: ");
```

```
        int first = s.nextInt();
        System.out.println("Enter second number: ");
        int second = s.nextInt();
        double quotient = (double)first / second;
        String result = String.format("%6.2f", quotient);
        System.out.println("Quotient is: " + result);
        s.close();
    }
}
```

In the statement `String.format("%6.2f", quotient);` the values inside the `"%f"` are `6` and `2`. They define how the floating-point number will be formatted. The `6` specifies the minimum width of the output, including the digits, the decimal point, and any padding spaces. The `2` indicates the number of digits to display after the decimal point. So, the format ensures the output aligns neatly, especially when printing multiple numbers in a column.

Here's the sample output:

```
java DisplayAverage
Enter first number:
10
Enter second number:
3
Quotient is:   3.33
```

7.7 The `File` Class and the `File` Object

In Java, the `File` class is a versatile tool that represents both files and directories, making it easier to manage and manipulate them. It doesn't directly read or write data; instead, it provides methods to check file properties, create or delete files, and navigate directory structures. You can create a `File` object using a simple file name, a directory path with a file name, or even by combining an existing `File` object representing a directory with a file name.

For example:

```
// creates a reference to a file
// in the current working directory
 File myFile = new File("myfile.txt");
```

Similarly:

```
//  represents a file inside the "MyDocs" folder.
File myFile = new File("MyDocs", "myfile.txt");
```

When dealing with directories, the `File` class treats them similarly to files, allowing you to navigate or locate other files within them.

Using `File myDir = new File("MyDocs");` followed by `File myFile = new File(myDir, "myfile.txt");` makes it easier to manage files within a specific directory.

Additionally, the `File` class offers methods like `exists()`, `canRead()`, and `canWrite()` to check file properties and ensure safe file operations.

7.7.1 The Relationship Between the `File` Object and the Physical File on Your Hard Drive

Once a `File` object has been instantiated, a link of that `File` object to the physical file you have on your hard drive will be established. The `File` object serves as a mirror to your physical file.

Let's say you have this:

```
File myFile = new File("myfile.txt");
```

A `File` object will then be created in the heap memory, linking itself to your file `myfile.txt` on your hard drive. You can then gain access to that file object by its reference variable `myFile`.

Figure 7.3 Stack-heap diagram for file objects

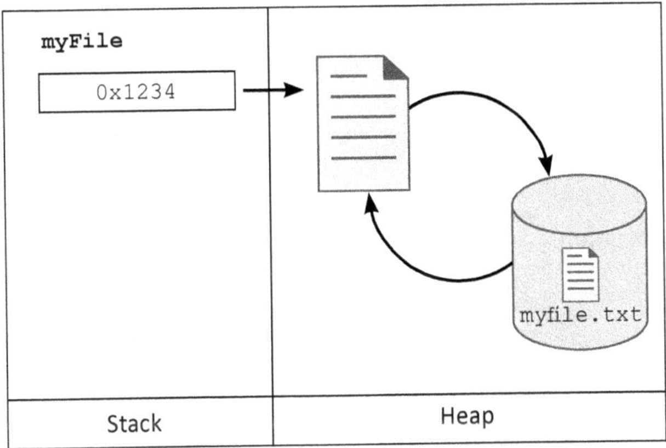

7.7.2 Common `File` Tests and Utilities Methods

Under the `java.io.File` class, there are helper methods that can retrieve information, test the file, or even check whether the file was modified. Here are some of the helper methods under the `File` class.

Table 7.4 Common `File` Methods

Category	Method Name	Purpose
File Information	`String getName()`	Retrieves the name of the file or directory.
	`String getPath()`	Returns the file path as a string.
	`String getAbsolutePath()`	Provides the complete absolute file path.
	`String getParent()`	Returns the parent directory of the file or directory.
	`long lastModified()`	Shows the last modified timestamp of the file.
	`long length()`	Displays the size of the file in bytes.
File Modification	`boolean renameTo(File newName)`	Renames the file to the specified new name.
	`boolean delete()`	Deletes the file or directory.
Directory Utilities	`boolean mkdir()`	Creates a new directory.
	`String[] list()`	Lists the names of files and directories within a directory.
File Tests	`boolean exists()`	Checks if the file or directory exists.
	`boolean canWrite()`	Verifies if the file is writable.
	`boolean canRead()`	Verifies if the file is readable.
	`boolean isFile()`	Tests whether the file denoted is a regular file or not.

Category	Method Name	Purpose
	`boolean isDirectory()`	Checks if the path is a directory.
	`boolean isAbsolute()`	Confirms if the path is absolute.
	`boolean isHidden()`	Checks if the file or directory is hidden.

7.8 Reading Inputs From a File

So far, all the input has been done from the keyboard, which happens to be the standard input device. This allows the program to read data directly from a user. However, there are cases in which programs are required to receive input from a file. Reading and writing data to a file is possible in application programs.

The class that is responsible for reading data from a file is the `FileReader` class:

```
BufferedReader br = new BufferedReader(new FileReader(myFile));
```

The `br` is a reference name variable, while `myFile` is the name of the file where data will be read.

Java `File` Input Programs:

```
// ReadFile.java
import java.io.*;
public class ReadFile {
  public static void main(String[] args) {
    File myFile = new File(args[0]);
    try {
      BufferedReader br = new
          BufferedReader(new FileReader(myFile));
```

```
         String str;
         while ((str = br.readLine())!= null) {
            System.out.println("Read: " + str);
         }
         br.close();
      }
      catch (FileNotFoundException fnfe) {
         System.out.println("File Not Found "+ myFile);
      }
      catch (IOException ioe) {
         System.out.println("IOException");
      }
   }
```

The given program will ask the user to input the name of the file in the command line (`args[0]`), and the program will open the file. If the given file is present, it will read each line of text and display all the contents of the file on the screen. If the file is not present, a `FileNotFoundException` will be thrown.

7.9 Writing Inputs to a File

To enable the write capability to a file, we use the `FileWriter` class and the `PrintWriter` class.

```
PrintWriter out = new
     PrintWriter(new FileWriter(myFile));
```

- `out` is a reference variable that will display the stream to the specified file.
- `myFile` is the name of the file where the input `String` will be stored.

Java File Output Programs:

```
// WriteFile.java
import java.io.*;
public class WriteFile {
  public static void main(String[] args) {
    File myFile = new File(args[0]);
    try {
      BufferedReader bf = new BufferedReader(new
               InputStreamReader (System.in));
     // this allows the actual writing to the file.
      PrintWriter out =
         new PrintWriter(new FileWriter(myFile));

      String s;
      System.out.println("Enter File Text: ");
      System.out.println("Press ctrl + Z to end! ");
      while ((s = bf.readLine()) != null) {
         out.println(s);
      }
      bf.close();
      out.close();
    }
    catch (FileNotFoundException fnfe) {
        System.out.println("File Not Found " + args[0]);
    }
    catch (IOException ioe)  {
        System.out.println("IOException occurred");
    }
  }
}
```

In the given code, a `BufferedReader` object was created to be able to get keyboard inputs from the user, but instead of writing it to the screen, it was written to the file by using the `PrintWriter` object.

Optionally, if you used: `PrintWriter out = new PrintWriter(new FileWriter(myFile, true));` with `true` as the append value, it will allow you to write appended pre-written text to your file.

Chapter Summary

- Java I/O streams provide a standard way to read from and write to various data sources, such as files, network connections, or memory buffers. They are categorized into byte streams (for handling raw binary data) and character streams (for handling text data), allowing efficient and flexible input/output operations in Java.
- Fundamental stream classes in Java include `InputStream` and `OutputStream` for handling byte-based data, and `Reader` and `Writer` for handling character-based data. These classes serve as the foundation for more specialized stream classes, providing the basic framework for reading and writing data in Java applications.
- Data within streams in Java is processed sequentially, where byte streams handle raw binary data, and character streams manage text data. Streams can either be input streams, which read data from a source, or output streams, which write data to a destination, ensuring a smooth flow of data between the program and external resources.
- I/O stream chaining in Java refers to the practice of connecting multiple streams together to perform complex data processing tasks. By chaining streams, you can combine functionalities, such as reading from a file while buffering the input or compressing data before writing it, creating a more efficient and flexible way to handle I/O operations.
- The `BufferedReader` and `InputStreamReader` classes in Java work together to efficiently read text data from an input stream. `InputStreamReader` converts byte streams to character streams, while `BufferedReader` adds buffering capabilities, allowing for more efficient reading of text, especially when dealing with large amounts of data or reading line-by-line.

- The `Scanner` class in Java is a versatile utility for parsing and reading input from various sources, such as user input; it can also be used for file IO. It provides methods to easily parse and retrieve different data types, like integers, strings, and doubles, making it a convenient tool for handling input in Java programs.
- The `String.format()` method in Java formats strings by returning a formatted string based on a specified format and arguments, without modifying the original string. The `System.out.printf()` method also formats strings similarly but directly prints the formatted output to the console, combining formatting and output in a single step.
- To read and write to files in Java, you typically use classes from the `java.io` package, such as `FileReader` and `FileWriter` for character-based data, and `FileInputStream` and `FileOutputStream` for byte-based data. These classes allow you to open files for reading or writing, and when used in combination with buffering classes like `BufferedReader` and `BufferedWriter`, they can efficiently handle large files by reducing the number of I/O operations.

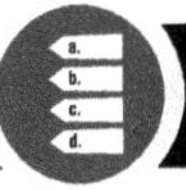

Multiple Choice Questions

1. **By default, the `readLine()` method of the BufferedReader class can accept ________ values.**
   a. `int`
   b. `boolean`
   c. `char`
   d. `String`

2. **The ________ package contains many classes that programs can use to read and write data from different data sources, like the keyboard and files.**
   a. `java.util`
   b. `java.lang`
   c. `java.awt`
   d. `java.io`

3. **The ________ class represents a physical file or folder (directory) on your hard drive.**
   a. directory
   b. subfolder
   c. `File`
   d. `package`

4. **Java _______ streams are used to perform input and output of an 8-bit stream**
   a. character
   b. `byte`
   c. bit
   d. Unicode

5. **Java ______ streams are used to perform input and output for a 16-bit stream.**

   a. bit
   b. character
   c. ASCII
   d. `byte`

6. **Which import statement MUST be present if you are going to use the Scanner class in your code?**

   a. `import java.util.Scanner;`
   b. `import java.lang.Scanner;`
   c. `import java.awt.Scanner;`
   d. `import java.io.Scanner;`

7. **When accepting a user input using the `Scanner` class, what type of `Exception` object will be thrown if the user does not enter the correct value for a variable's data type?**

   a. `InputMismatchException`
   b. `NumberFormatException`
   c. `IOException`
   d. `IllegalArgumentException`

8. **What package should you import if you plan to use both the BufferedReader and InputStreamReader?**

   a. `java.io` package
   b. `java.lang` package
   c. `java.awt` package
   d. `java.util` package

9. **The InputStreamReader class has a method called _______ that will release any system resources associated with it.**

   a. `release()`
   b. `disconnect()`
   c. `close()`
   d. `stop()`

10. **Which class is used under the `java.io` package that is used to gain read access to a text file?**

    a. `FileInputStream`
    b. `File`
    c. `FileReader`
    d. `FileInputStreamReader`

Multiple Choice Answer Key

1 – d	2 – d	3 – c	4 – b	5 – b
6 – a	7 – a	8 – a	9 – c	10 – c

Coding Task 1: Usernames and Passwords

Create a Java Program that will read a text file named: `userList.txt`. The filename should be passed via the command line. It contains a list of usernames and passwords. The odd-numbered lines contain the usernames, while the even-numbered lines contain the passwords for the usernames in the previous lines. Here's a sample content for the list of `Strings` in the `userList.txt`:

```
user1
password1
user2
password2
user3
password3
```

Once the `String` values are read in the file `userList.txt`, you are going to create two text files named: `usernames.txt` and `password.txt`. The file `usernames.txt` will contain all the usernames located in the odd-numbered lines, while the passwords will be written in the `password.txt` file. No output message is required, but it will be nice to give a confirmation that your code worked. The only way you'll know if your code works is when both `usernames.txt` and `password.txt` contain the correct `String` values.

```
usernames.txt should contain the following:
user1
user2
user3
```

```
password.txt should contain the following:
password1
password2
password3
```

Coding Task 2: File IO in Java

Objective:

Write a Java program that reads data from one text file, processes the data, and writes the processed data to another text file.

Exercise Details:

Input File:

You will be given a text file named `input.txt` that contains multiple lines of text. Each line consists of a person's name followed by their age, separated by a comma and a space. Assume that the maximum number of names in the list is 10. In this example, there are only 4 names in the list:

```
Lawrence Decamora, 28

Michelle Crisostomo, 34

Danielle Lauren, 23

Elisha Bernice, 42
```

Task 1: Reading the File

Read the content of the input.txt file using a `BufferedReader` or `FileReader`.

Store each line in an array of `String` for further processing.

Task 2: Processing the Data

For each line in the list, extract the person's name and age.

Convert the age to an integer and increase it by 1 (simulating a birthday).

Task 3: Writing to a New File

Write the updated information (name and new age) to a new file named `output.txt` using a `BufferedWriter` or `FileWriter`. The format of each line in the `output.txt` file should be the same as the input file, but with the updated ages. For example:

```
Lawrence Decamora, 29
Michelle Crisostomo, 35
Danielle Lauren, 24
Elisha Bernice, 43
```

Task 4: Handle Exceptions

Ensure that your program handles exceptions such as `FileNotFoundException`, `IOException`, etc. If the input file is missing, display an appropriate error message to the user.

Task 5: Close Resources

Ensure that all file streams are properly closed after reading and writing operations to prevent resource leaks.

Task 6: Displaying the contents of the file

Display the contents of both the input file (`input.txt`) and the output file (`output.txt`) in the console.

CHAPTER 8

Introduction to Java Collection and the Generics Framework

Key Learning Objectives

- Understand the role of the Collection interface in Java
- Distinguish the difference between Set, List, and Map and their use.
- Learn how to iterate over collection elements using the Iterator interface
- Become comfortable with Generics and the benefits of type-safe collections
- Utilize the diamond operator (<>) to simplify code when instantiating collection classes
- Develop skills to choose the appropriate collection for a given problem

This chapter introduces you to the powerful Java Collections Framework, a set of classes and interfaces designed to efficiently store, retrieve, and manipulate groups of objects. Here, you'll learn about the main collection types — `Set`, `List`, and `Map` — and how each is used to solve different kinds of problems. You'll also discover the `Iterator` interface, a convenient way to iterate through elements in a collection. Furthermore, this chapter covers generics, a key feature that lets you create type-safe collections, reducing bugs and making your code more reliable and easier to maintain. Lastly, you'll learn about the diamond operator (<>), a helpful shortcut for instantiating collection classes without needing to repeat the *type* parameters. By the end of this chapter, you will be well-equipped to leverage Java's collection framework to store and manipulate data effectively in your own applications.

8.1 The `Collection` Interface

The `Collection` interface is the root interface in the collection hierarchy. A collection represents a group of objects known as its elements.

Some collections allow duplicate elements, and others do not. Some are ordered and others are unordered. The Java development kit does not provide any direct implementation of this interface. It only provides implementation of more specific sub-interfaces like the set collection and the list collection. This interface is typically used to pass collections around and manipulate them where maximum generality is desired. Now, if you look at this hierarchy, you will see that the collection interface is at the top, and some of the known subclasses or sub-interfaces are set and list.

To explore the full spectrum of sub-interfaces under the `Collection` interface, the most reliable resource is the official Java API documentation. Within this documentation, navigate to the `java.util` package, where you will find all interfaces pertinent to the `Collection` framework. Focusing on the `Collection` interface, you'll observe several direct sub-interfaces listed.

Key among these are `Set` and `List`, both of which are prominent sub-interfaces of `Collection`. These interfaces inherit the method signatures defined by `Collection`. For instance, the `add()` method, shared by both `Set` and `List`, allows you to insert elements into the collection. The method returns a boolean value—`true` if the addition is successful, and `false` if the element cannot be added.

Figure 8.1 Collection Interface

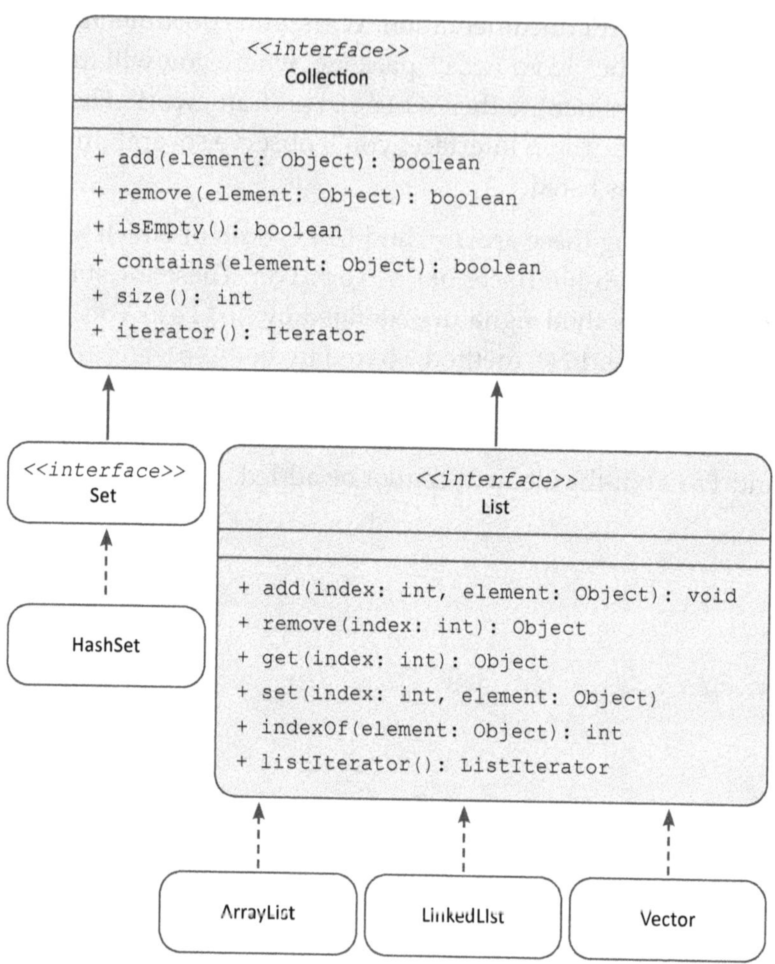

However, `List` distinguishes itself by offering overloaded variants of the `add()` method, one of which accepts two arguments. This provides greater flexibility in how elements are added to a list, illustrating the nuanced differences between these interfaces within the collection hierarchy.

We can also remove elements from a collection. However, `List` provides the additional capability of removing an element based on its index. It's important to note that sets maintain uniqueness—every element in a `Set` must be unique, and the order of elements is not guaranteed. Conversely, `List` allows duplicate elements and maintains a specific order.

When considering the implementing classes of these interfaces, `Set` has prominent implementations like `HashSet` and `TreeSet`. For `List`, notable implementations include `Vector`, `ArrayList`, and `LinkedList`.

Returning to the fundamental concept of what a `Collection` is, it represents a group of objects referred to as its elements. This concept is somewhat analogous to arrays, which are also objects representing a group of elements. However, the distinction lies in the flexibility and type of elements they can contain. A `Collection` can only hold objects, while an array can hold either a group of objects or a group of primitive values. For example, you can have an array of objects, such as an array of `String` or Student objects, or an array of primitive values, such as `int` or `boolean`. To convert an array of primitives to a collection of objects, the wrapper classes come into play.

Another key difference is in their memory management and thread safety. Modern collections are not automatically thread-safe, though some, like synchronized lists, provide thread safety. Thread-safe collections tend to be heavier in terms of memory usage compared to arrays, which are not thread-safe and therefore lighter. This thread safety allows collections to be shared by multiple concurrent users, whereas arrays are typically intended for single-user scenarios.

Consider the example of a social media platform like Facebook: you would refer to your friends as a collection rather than an array of friends. Referring to friends as a collection rather than an array is more appropriate because collections, such as `List` or `Set` in Java, provide flexible and dynamic storage that allows easy addition or removal of friends without predefined size limitations. Unlike arrays, collections are better suited for handling real-world data, where the number of elements can vary and operations like searching or sorting are commonly required.

Every time you approve a friend request, you add them to your collection of friends. The size of your friends' collection increases dynamically. In the same way, when you unfriend someone from your collection, you remove them from your collection, and the number of your friends drops dynamically every time you unfriend someone. On the contrary, the size of an array is always fixed upon creation; its size cannot change dynamically.

Additionally, collections are mutable; they grow as you add more elements and shrink as you remove them. Arrays, in contrast, are fixed in length—once an array is created, its size remains fixed for the duration of its lifecycle.

Table 8.1 lists the key differences between Collections and Arrays.

Table 8.1 Collections vs. Arrays

Collections	Arrays
A **single object** that represents a group of *objects*, known as its elements.	A **single object** that represents a group of *objects* or *primitive values*.
Some are thread-safe, heavyweight if synchronized.	NOT thread-safe, lightweight.

Collections	Arrays
Mutable, the Collection object can grow larger or smaller depending on whether you're adding more objects or removing objects from inside the collection.	The length is fixed upon the creation of the array object.
Uses the method `.size()` to check the number of elements inside the Collection.	Uses the attribute `.length` to check the size of the array.

8.2 The `Set` Interface

Let's delve into one of the subinterfaces of the `Collection` interface, known as the `Set` interface. The `Set` interface is designed to store a collection of unordered elements, meaning that the order of elements is not guaranteed.

One of the fundamental characteristics of a `Set` is its prohibition of duplicate elements. This implies that every element within a `Set` must be unique. If you attempt to add a duplicate element, only the first instance will be retained, while subsequent duplicates will be ignored.

Moreover, a `Set` permits at most one `null` element. To illustrate this, consider the sample code provided below. Here, we declare a `Set` object, instantiated with a `HashSet`, which is one of the most commonly used implementations of the `Set` interface. From lines 6 to 12, various objects are added to the `Set`, including some that are duplicates. These duplicate objects do not trigger any runtime errors. Instead, only the first occurrence of each object is stored in the `Set`. When the `Set` is printed, only the unique elements are displayed, as duplicates are automatically excluded.

Here's a sample saved as `SetExample.java`.

```
import java.util.*;
public class SetExample {
 public static void main(String[] args) {
     Set set = new HashSet();

     set.add("Lawrence");
     set.add(100);
     set.add(false);
     set.add("java");
     set.add("java");
     set.add("Lawrence");
     set.add(3.0);

     System.out.println(set);
  }
}
```

If you compile the given code directly using the javac command, you will get a notification:

```
javac SetExample.java <ENTER>
Note: SetExample.java uses unchecked or unsafe operations.
Note: Recompile with -Xlint:unchecked for details.
```

To use an option to display the compilation warning, we use this command:

```
javac -Xlint:unchecked SetExample.java <ENTER>
```

Using the `-Xlint:unchecked` compilation options will display the following compilation warnings:

```
SSetExample.java:6: warning: [unchecked] unchecked call
to add(E) as a member of the raw type Set
    set.add("Lawrence");
           ^
  where E is a type-variable:
    E extends Object declared in interface Set
SetExample.java:7: warning: [unchecked] unchecked call
to add(E) as a member of the raw type Set
    set.add(100);
           ^
  where E is a type-variable:
    E extends Object declared in interface Set
SetExample.java:8: warning: [unchecked] unchecked call
to add(E) as a member of the raw type Set
    set.add(false);
           ^
  where E is a type-variable:
    E extends Object declared in interface Set
SetExample.java:9: warning: [unchecked] unchecked call
to add(E) as a member of the raw type Set
    set.add("java");
           ^
  where E is a type-variable:
    E extends Object declared in interface Set
SetExample.java:10: warning: [unchecked] unchecked call
to add(E) as a member of the raw type Set
    set.add("java");
           ^
  where E is a type-variable:
    E extends Object declared in interface Set
SetExample.java:11: warning: [unchecked] unchecked call
to add(E) as a member of the raw type Set
    set.add("Lawrence");
           ^
```

```
   where E is a type-variable:
     E extends Object declared in interface Set
 SetExample.java:12: warning: [unchecked] unchecked call
 to add(E) as a member of the raw type Set
     set.add(3.0);
            ^
   where E is a type-variable:
     E extends Object declared in interface Set
 7 warnings
```

Warnings are just warnings; they are not errors. Therefore, we can still run them using the command:

```
java SetExample <ENTER>
```

Runtime Output:

```
[java, 100, false, Lawrence, 3.0]
```

After compiling and running `SetExample.java`, we notice a compilation warning, which is common for the initial examples because *Java Generics* have not been used yet. Although these warnings are not errors, it is generally good practice to treat them seriously.

In this example, although seven objects were added to the `Set`, only five unique elements are present when the `Set` is printed. Notice that both the `"java"` and `"Lawrence"` strings appear only once, despite being added twice, in lines 9 and 10 for `"java"` and lines 6 and 11 for the string `"Lawrence"`. A `HashSet` in Java is a collection that stores unique elements and provides efficient operations for adding, removing, and searching items using a hash table. It does not maintain any specific order of elements and automatically eliminates duplicates. This behavior is due to the add() method inherited from the Collection interface, which returns a boolean value. If an attempt is made to add a duplicate element, the method simply returns false, indicating that the element was not added.

8.3 The List Interface

Now let's examine the List interface and understand how it differs from the Set interface. Unlike a Set, a List represents an ordered collection, meaning that the sequence in which elements are added is preserved. The List interface provides the user with precise control over where each element is inserted. You can access elements by their index positions and even search for specific elements within the list. Additionally, the List interface allows for duplicate elements, making it possible to have repeated entries. Here's a sample code saved as: ListExample.java.

```
import java.util.*;
public class ListExample {
 public static void main(String[] args) {
   List list = new ArrayList();

   list.add("Lawrence");
   list.add(100);
   list.add(false);
   list.add("java");
   list.add("java");
   list.add("Lawrence");
   list.add(3.0);

   System.out.println(list);

  }
}
```

Runtime Output:

```
[Lawrence, 100, false, java, java, Lawrence, 3.0]
```

Just like the code in the previous example, you will encounter a similar compilation warning. We'll get rid of that warning in the later part of this chapter when we discuss the concept of Java Generics.

Consider the code segment on line 4, where we declare a `List` object using the `ArrayList` implementation. There are several `List` implementations available, including `ArrayList`, `Vector`, and `LinkedList`. The `ArrayList` is similar to what arrays are. In lines 6 through 12, we add several objects to the list—seven in total, including some duplicates.

These duplicates are accepted into the list without any issues, and when we print the list on line 14, the output shows that the order in which the elements were added is maintained. The first object added is the string `"Lawrence"`, and the last is the `double` value `3.0`, which is reflected in the output.

To better understand the differences between `Set` and `List`, let's look at a code example involving a `Set`, specifically a `HashSet`, which is one of the class implementations of `Set`. We add several objects to the set and print the result.

When the `SetExample` code is executed, only five objects are printed, even though seven were added. This is because the `Set` interface, as discussed previously, does not allow duplicates. The `add()` method returns `false` when a duplicate is detected, and thus, those elements are not added again. Additionally, the order of elements in a `Set` is not preserved, as evidenced by the fact that "Lawrence" was added first, but "java" is retrieved first.

In contrast, when we use a List, specifically an ArrayList, with the same set of objects, the output reflects all seven elements in the order they were added. After compiling and running ListExample.java, we observe that the list maintains the order of elements, as expected.

So, when should you use a Set, and when should you use a List?

If the arrangement of elements is not important, and you want to ensure that all elements are unique, a Set is appropriate. For example, if you're keeping track of a group of classmates, where the order in which they enter a room does not need to be preserved, a Set would be suitable.

On the other hand, if the order of elements matters, a List is a better choice. For example, when making a grocery list where the order of items may reflect priorities, such as buying food items before laundry items, a List is ideal, and duplicates are allowed.

To summarize the difference between a Set and a List, you can check out Table 8.2.

Table 8.2 Comparison between Set and List Interfaces.

Aspect	Set	List
Order	Unordered elements	Ordered elements
Duplicates	Not allowed (each element must be unique)	Allowed (you can have multiple copies of the same element)
Index Access	Not supported (can't retrieve by index)	Supported (you can retrieve, insert, or remove by index)

Aspect	`Set`	`List`
Implementation Examples	`HashSet`, `TreeSet`	`ArrayList`, `LinkedList`, `Vector`
Primary Use	To keep unique elements	To preserve order or allow duplicates
Performance	Fast for the search and removal of elements	Fast for positional operations, accessing by index

8.4 The `Map` Interface

A `Map` is a type of data structure used to store elements that can be accessed via a unique identifier known as a key. This structure consists of pairs, where each key (denoted as `K`) is associated with a value (denoted as `V`).

When examining the Java API, particularly the `java.util` package, you'll find the `Set` and `List` interfaces, each marked with the diamond operator (<>) containing the element type (`E`). This indicates that when creating a `Set` or `List` object, you need to specify the type of elements it will hold. In contrast, for `Map`, you'll see `K` and `V`, indicating that a `Map` associates keys with values.

Let's look at an example. In line 4 of the code, a `Map` object is declared and implemented as a `HashMap`. `HashMap` is one of the implementing classes of the `Map` interface.

A `HashMap` in Java is a collection that stores key-value pairs, allowing fast retrieval, insertion, and deletion of values based on their associated keys. It uses a hash table for efficient data management and does not guarantee any specific order of the entries. A hash table is like a magical filing cabinet where each drawer has a special label called a "*key*." When you want to put something in the cabinet, you use a special rule (called a hash function) to decide which drawer it belongs in. Later, if you want to find it, you just use the same label to quickly open the right drawer without checking all of them. It's a super-fast way to keep things organized and easy to find. In the code below, we will demonstrate the use of `Map` and `HashMap`.

The given file is saved as: `MapExample.java`.

```
import java.util.*;
public class MapExample {
public static void main(String[] args) {
  Map map = new HashMap();
  map.put("one", "uno");
  map.put("two", "dos");
  map.put("three","tres");

  System.out.println("Original Map Values: ");
  System.out.println(map);

  // Overwrites the previous assignment
  map.put("three","trois");

  System.out.println("Updated Map Values: ");
  System.out.println(map);


  // Returns set view of keys
  Set set1 = map.keySet();
  System.out.println(set1);

  // Returns Collection view of values
```

```
        Collection collection = map.values();
        System.out.println(collection);

        // Returns set view of key value mappings
        Set set2 = map.entrySet();
        System.out.println(set2);

        System.out.println(set1 + "\n" +
                    collection + "\n" + set2);
    }
}
```

Runtime Output:

```
Original Map Values:
{one=uno, two=dos, three=tres}
Updated Map Values:
{one=uno, two=dos, three=trois}
[one, two, three]
[uno, dos, trois]
[one=uno, two=dos, three=trois]
[one, two, three]
[uno, dos, trois]
[one=uno, two=dos, three=trois]
```

Unlike `Set` or `List`, where you would use the `add()` method to insert elements, with `Map`, you use the `put()` method. This method inserts a pair of objects—one serving as the *key* and the other as the *value*. In this example, the *keys* and *values* are both strings.

For instance, the key `"one"` is associated with the value `"uno"`, `"two"` with `"dos"`, and `"three"` with `"tres"`.

When you print the original Map values using System.out.println(map);, you will see the *key-value* pairs displayed in the runtime output.

However, if you insert another *key-value* pair with an existing *key*, like "three" with a new *value* "teen", it will override the previous value associated with that *key*. This means the pair "three", "teen" will replace the pair "three", "tres" in the map.

After compiling and running MapExample.java, despite the warning about not using the diamond operator or generics, the original map will display {one=uno, two=dos, three=tres}, and the updated map will display {one=uno, two=dos, three=teen}.

It's important to note that key objects in a Map must be unique. To retrieve *keys*, the keySet() method is used, which returns a set of all keys in the map. Since both *keys* and *values* in a Map are objects, primitive types cannot be added directly. However, starting from Java 5, autoboxing allows primitive values to be automatically converted to their wrapper class equivalents.

When working with a Map, each entry (*key-value* pair) is managed as an entry set, which can be retrieved using the entrySet() method. Printing objects such as a Map, Set, or List typically displays elements within square brackets ([]).

If you want to customize the way you print these elements—for example, printing just the keys in a specific format—you can iterate over the collection. Iteration can be performed in two ways: forward or backward, depending on the desired outcome. For us to do that, we have to discuss the concepts of the Iterator interface.

8.5 The `Iterator` Interface

We've discussed the `Set`, `List`, and `Map` interfaces, and how to add and print objects from these collections using methods like `System.out.println()`. When printing these objects, they are usually displayed within square brackets (`[]`).

However, if you want to format the output differently, or if you need to perform computations or store the values elsewhere while iterating through the collection, you can use the `Iterator` or `ListIterator` interfaces.

Figure 8.2 *`Iterator`* **and** *`ListIterator`* **interfaces**

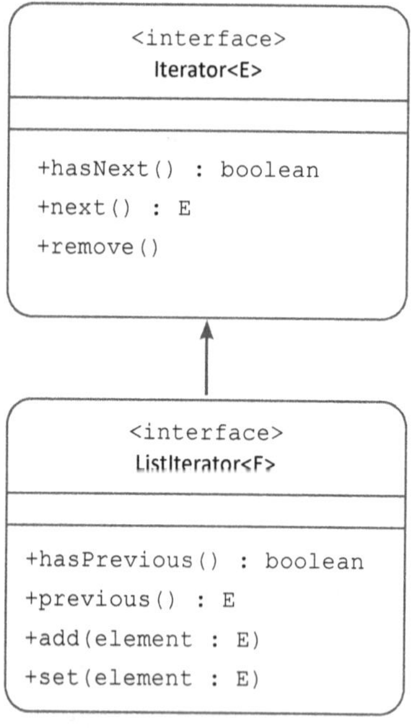

An `Iterator` is an object used for looping through collections, such as an `ArrayList` or a `HashSet`. The term "iterating" refers to looping, and iterators support forward iteration only. This makes the `Iterator` interface ideal for use with `Set` collections, which do not maintain any specific order.

For `List` collections, the `ListIterator` interface can be used. Lists maintain the order of elements, so the `ListIterator` allows you to iterate both forward and backward through the list.

Let's consider an example. In our code, we have a `Person` class that we've used in previous examples. This class has two attributes: `name` and `age`. It also includes a constructor, getter methods (`getName()` and `getAge()`), a `toString()` method, an `equals()` method, and a `hashCode()` method.

```
public class Person {
  private String name;
  private int age;

  public Person(String name, int age) {
      this.name = name;
      this.age = age;
  }

  // getters
  public String getName() {
      return name;
  }
  public int getAge() {
      return age;
  }

  // toString method
  public String toString() {
```

```
        return getName() + ", at age " + getAge();
    }

    // equals and hashCode method
    public boolean equals(Object obj) {
        boolean result = false;

        if (obj != null && obj instanceof Person ) {
            Person p = (Person) obj;
            if (this.getName().equals(p.getName())
             && this.getAge() == p.getAge()) {
                    result = true;
            }
        }
        return result;
    }

    public int hashCode() {
        return getName().hashCode() ^ getAge();
    }
}
```

In our test class, we've instantiated five `Person` objects. We'll create a collection called `myFriends`, implemented as a `HashSet`, and add these `Person` objects to it. Since `HashSet` does not guarantee order, when we print the collection using `System.out.println(myFriends);`, the order of the elements may differ from the order in which they were added. Additionally, the output will be enclosed in square brackets. Here's the code for our test class, `TestPerson.java`.

```
import java.util.*;

public class TestPerson {
 public static void main(String[] args) {
       int sum = 0;

       Person p1 = new Person("Edwin", 20);
       Person p2 = new Person("Percy", 30);
       Person p3 = new Person("Ericson", 24);
       Person p4 = new Person("Jerry", 18);
       Person p5 = new Person("Lesley", 27);

       Set myFriends = new HashSet();

       myFriends.add(p1);
       myFriends.add(p2);
       myFriends.add(p3);
       myFriends.add(p4);
       myFriends.add(p5);

       System.out.println(myFriends);

  }
}
```

Runtime Output:

```
[Edwin, at age 20, Lesley, at age 27, Percy, at age 30,
Jerry, at age 18, Ericson, at age 24]
```

If we want to print each friend's information individually and, say, calculate the average age of the group, we can use the `Iterator` interface. Since we're working with a `Set`, we can use an iterator.

Here's how it works:

We create an iterator using `myFriends.iterator()`. The `iterator()` method is defined in the `Set` interface and returns an iterator object.

The iterator provides three main methods: `hasNext()`, `next()`, and `remove()`. These methods help us loop through the collection.

We use a while loop to iterate over the set. The condition `it.hasNext()` checks if there are more elements to process. If `true`, we retrieve the next element with `it.next()`.

If needed, we can store the retrieved element in a variable, such as a `Person` object, by casting it appropriately.

To calculate the average `age`, we compute the `sum` of the ages of all the `Person` objects using `p.getAge()`, then divide by the total number of elements, obtained with `myFriends.size()`. Below is the updated version of our test class, `TestPerson.java`, this time using the `Iterator` object.

```
import java.util.*;

public class TestPerson {
 public static void main(String[] args) {
      int sum = 0;

      Person p1 = new Person("Edwin", 20);
      Person p2 = new Person("Percy", 30);
      Person p3 = new Person("Ericson", 24);
```

```
        Person p4 = new Person("Jerry", 18);
        Person p5 = new Person("Lesley", 27);

        Set myFriends = new HashSet();

        myFriends.add(p1);
        myFriends.add(p2);
        myFriends.add(p3);
        myFriends.add(p4);
        myFriends.add(p5);

        // System.out.println(myFriends);

        Iterator it = myFriends.iterator();

        while(it.hasNext()) {

            Person p = (Person) it.next();
            System.out.println(p);
            sum += p.getAge();
        }

        System.out.println("Sum: " + sum);
        double averageAge = (double)sum/myFriends.size();
        System.out.println("Average age of my Friends: "
                                + averageAge);

    }
}
```

Finally, we compile and run the revised `TestPerson.java`. The output will show each friend's details, and the average `age` will be displayed.

Runtime Output:

```
Edwin, at age 20
Lesley, at age 27
Percy, at age 30
Jerry, at age 18
Ericson, at age 24
Sum: 119
Average age of my Friends: 23.8
```

This approach allows us to iterate through each element in the collection, perform computations like averaging, and display the results in a customized format.

8.6 An Introduction to the Generics Framework

When we compile our Java source code, we often see compilation warnings like this one when we compile our `MapExample.java` file.

```
Note: MapExample.java uses unchecked or unsafe operations.
Note: Recompile with -Xlint:unchecked for details.
```

These warnings occur because we might be using the collections framework without generics. So, what exactly is the generics framework, and how can we use it to eliminate these warnings?

The generics framework was introduced in Java 5, and it provides a way to parameterize types, known as elements. If you look at the Java API documentation for any collection interface or class, you'll notice a diamond operator (<>) with a type parameter, like E (for elements). This is also true for sub-interfaces of collections. For Map, instead of E, we use K and V for key-value pairs.

Generics reduce the amount of code required when adding or retrieving objects from a collection. For instance, in previous examples, when we used an iterator to retrieve objects from a collection, we had to cast the returned object because the next() method returns a generic Object type.

By using generics, we introduce compile-time type safety, allowing the compiler to catch invalid types during compilation. This also eliminates the need for explicit casting. Let's see how this works with a new version of our loop, using generics. Refer to the code below, lines 26 to 31.

You can declare generics in your collection declarations. For example, in a Set declaration, you can specify that it only accepts Person objects using the diamond operator (<>). Both the left-hand and right-hand sides of the declaration should specify the Person type. This ensures that only Person objects can be added to the Set. If you attempt to add a non-Person object, such as a String, it will result in a compilation error. This error helps ensure that only the correct types are added to the collection, and it appears during compile-time, making it easier to debug. Look at the commented line 20 in the code below; using that line will cause an error.

Let's revisit our test class, TestPerson.java, but this time, it uses the diamond operator (<>).

```
1   import java.util.*;
2
3   public class TestPerson {
4   public static void main(String[] args) {
5      int sum = 0;
```

```
        Person p1 = new Person("Edwin", 20);
        Person p2 = new Person("Percy", 30);
        Person p3 = new Person("Ericson", 24);
        Person p4 = new Person("Jerry", 18);
        Person p5 = new Person("Lesley", 27);

        Set<Person> myFriends = new HashSet<Person>();

        myFriends.add(p1);
        myFriends.add(p2);
        myFriends.add(p3);
        myFriends.add(p4);
        myFriends.add(p5);
//      myFriends.add("My other friend.."); //! error

//      System.out.println(myFriends);

        Iterator<Person> it = myFriends.iterator();

        while(it.hasNext()) {

            Person p = it.next();
            System.out.println(p);
            sum += p.getAge();
        }

        System.out.println("Sum: " + sum);
        double averageAge =
                    (double) sum / myFriends.size();
        System.out.println("Average age of my Friends: "
                    + averageAge);
    }
}
```

Additionally, you can use an enhanced for-loop (also known as a "for-each" loop) to iterate over the collection. The enhanced for-loop is designed specifically for use with collections and arrays. It consists of two parts: the first part is a temporary object (e.g., `Person` p), and the second part is the collection (myFriends). This loop automatically retrieves each Person object from the collection, making the code even more concise.

Try to use the code snippet below to use the enhanced for loop by changing lines 21 to 31 of the test class `TestPerson.java`. It should still work, and it should also give you the same output.

```
for (Person p : myFriends) {
    System.out.println(p);
    sum += p.getAge();
}
```

8.7 The Diamond (<>) Operator

In Java SE7, a new language feature called the diamond operator (<>) was introduced to improve the usability of the Generics framework. Prior to Java SE7, whenever you declared a collection, you had to specify the type of objects that the collection would hold, both on the left-hand side (the declaration) and the right-hand side (the instantiation).

For example, when declaring a `Set` or `List`, you needed to specify the type twice, as shown below:

```
Set<Person> mySet = new HashSet<Person>();
```

However, with the introduction of the diamond operator in Java SE7, this redundancy is no longer necessary. You can now use the diamond operator on the right-hand side during the instantiation, which allows the compiler to infer the type from the left-hand side. This feature is also applicable to `Map` objects.

In older versions of Java, you had to repeat the type declaration during both the declaration and instantiation of a `Map`, like this:

```
Map<String, Integer> myMap = new HashMap<String,
Integer>();
```

With Java SE7 and later, the diamond operator simplifies this process. Now, you only need to specify the type on the left-hand side during the declaration, and you can use the diamond operator on the right-hand side.

This approach is known as type inference, where the compiler infers the type from the left-hand side declaration. This reduces code redundancy and makes your code more concise.

Let's take a look at an example. Suppose we have the following code that uses the diamond operator:

```
Set<Person> myFriends = new HashSet<>();
```

Here, we declare that the `Set` will hold `Person` objects, but thanks to the diamond operator, we no longer need to repeat the `Person` type on the right-hand side. The compiler automatically understands that the `HashSet` will only contain `Person` objects.

If we save, compile, and run the code, we will get the same output as before, but with a cleaner and more modern syntax. The diamond operator is a small but powerful addition to Java that enhances the readability and maintainability of your code, especially when working with the Generics framework.

8.8 Factory Methods for Immutable Collections (JDK 9)

In JDK 9, the `List`, `Set`, and `Map` interfaces introduced new static factory methods `of()` and `copyOf()` to create immutable collections. These methods make it easier and more concise to create fixed collections.

For example:

```
List<String> names = List.of("Alice", "Bob", "Charlie");

Set<Integer> numbers = Set.of(1, 2, 3);

Map<Integer, String> idToName = Map.of(1, "Apple", 2,
"Banana");
```

The collections created this way are immutable and will throw `UnsupportedOperationException` if you attempt to modify them. This is especially useful for creating constants or passing data that should not be changed.

8.9 Revisiting Local Variable Type Inference using `var` (JDK 10)

Let us revisit the discussion on JDK 10's `var` keyword that was discussed in Chapter 1. The `var` keyword also allows the compiler to infer the variable type, improving code readability when used with generic collections. This is especially useful when the type is obvious from the right-hand side:

```
var names = List.of("Jane", "John", "Doe");

var idMap = new HashMap<Integer, String>();
```

However, developers should use `var` judiciously to maintain readability, especially in complex declarations.

8.10 Collectors.toUnmodifiableList/Set/Map (JDK 10)

In JDK 10, Java introduced new collector methods that make it easier to create read-only collections directly from streams. These are the `Collectors.toUnmodifiableList()`, `Collectors.toUnmodifiableSet()`, and `Collectors.toUnmodifiableMap()` methods.

Before this, you would often collect elements into a regular collection and then wrap it using `Collections.unmodifiableList()` or similar methods. Now, you can produce an unmodifiable collection in one step, making the code cleaner and less error-prone. An unmodifiable collection cannot be changed after it's created, any attempt to add, remove, or update an element will result in an `UnsupportedOperationException`. This is useful when you want to ensure that the result of a stream operation stays consistent and is not altered by mistake elsewhere in your program.

Here's sample code to demonstrate these new methods.

```
import java.util.List;
import java.util.Set;
import java.util.Map;
import java.util.stream.Collectors;
import java.util.stream.Stream;

public class UnmodifiableCollectorsExample {
  public static void main(String[] args) {
```

```
        // Unmodifiable List
        List<String> names =
           Stream.of("Alice", "Bob", "Charlie")
           .collect(Collectors.toUnmodifiableList());
        System.out.println("List: " + names);

        // Unmodifiable Set
        // duplicates are removed
        Set<Integer> numbers = Stream.of(1, 2, 3, 3)
           .collect(Collectors.toUnmodifiableSet());
        System.out.println("Set: " + numbers);

        // Unmodifiable Map
        Map<Integer, String> idNameMap =
           Stream.of("A", "B", "C")
           .collect(Collectors.toUnmodifiableMap(
           String::length, // key: length of string
           value -> value  // value: the string itself
        ));
        System.out.println("Map: " + idNameMap);

        // Trying to modify will throw an exception
        // Uncomment to see UnsupportedOperationException
        // names.add("David");
    }
}
```

In this example, the stream elements are collected directly into unmodifiable collections without any extra wrapping step. This not only reduces boilerplate code but also helps enforce immutability right at the point of collection, making your data safer from unintended modifications.

8.11 Sequenced Collections (JDK 21)

In JDK 21, Java introduced the Sequenced Collection interfaces, which provide a consistent way to work with collections that have a defined encounter order—meaning the elements are stored in a specific sequence. These new interfaces include `SequencedCollection`, `SequencedSet`, and `SequencedMap`. They extend the existing collection types and add methods to easily access the first and last elements, as well as to traverse the collection in reverse order. Before this feature, you had to rely on different collection-specific methods or manual iteration to perform these tasks, which could vary between types. With sequenced collections, you can write code that works the same way for lists, sets, or maps that maintain order, improving readability and consistency.

```
import java.util.SequencedCollection;
import java.util.ArrayList;

public class SequencedCollectionExample {
   public static void main(String[] args) {
      // Creating a sequenced collection
      // (ArrayList implements SequencedCollection)
      SequencedCollection<String> fruits
                            = new ArrayList<>();
      fruits.add("Apple");
      fruits.add("Banana");
      fruits.add("Cherry");

      // Accessing first and last elements directly
      System.out.println("First fruit: "
                                   + fruits.getFirst());
```

```
        System.out.println("Last fruit: "
                                    + fruits.getLast());
        // Iterating in reverse order
        System.out.println("Fruits in reverse:");
        for (String fruit : fruits.reversed()) {
            System.out.println(fruit);
        }
    }
}
```

Here's the output after execution:

```
First fruit: Apple
Last fruit: Cherry
Fruits in reverse:
Cherry
Banana
Apple
```

In this example, we use `getFirst()` and `getLast()` to quickly retrieve the first and last elements, and `reversed()` to loop through the list from end to start—all without writing extra logic. This feature makes working with ordered collections more straightforward and consistent across different collection types.

These enhancements introduced in newer JDK versions build upon the strong foundation of the Java Collections Framework and Generics. They promote immutability, conciseness, and safer type handling, enabling developers to write more expressive and maintainable code.

Chapter Summary

- The `Collection` interface is the root of the collection hierarchy in Java, representing a group of objects known as its elements and serving as the basis for more specific sub-interfaces like `Set` and `List`. While the Java Development Kit (JDK) does not provide direct implementations of this interface, it is widely used for passing and manipulating collections with maximum generality, with details available in the official Java API documentation.
- Collection objects in Java are part of the Collections Framework, offering dynamic data structures like lists, sets, and queues, which can grow or shrink in size and provide various methods for manipulating elements. Arrays, on the other hand, are fixed in size and store elements of a specific data type, offering fewer built-in methods for element manipulation compared to collections.
- The `Set` interface in Java is designed to store a collection of unordered, unique elements, prohibiting duplicates and allowing at most one null element. When duplicate elements are added to a `Set`, only the first occurrence is retained, and subsequent duplicates are ignored.
- The `List` interface represents an ordered collection where the sequence of elements is maintained, allowing duplicates and enabling access by index, unlike the `Set` interface, which stores unordered, unique elements and disallows duplicates. When comparing examples using `List` and `Set`, the `List` (e.g., `ArrayList`) preserves the insertion order and includes all elements, while the `Set` (e.g., `HashSet`) omits duplicates and does not guarantee order.
- A `Map` is another data structure that stores elements as *key-value* pairs, where each key (`K`) is associated with a corresponding value (`V`), allowing for efficient retrieval

based on the `key`. Unlike `Set` and List interfaces, which use a single type parameter (`E`) for elements, the `Map` interface uses two type parameters (`K` and `V`) to define the types of *keys* and *values* it holds.

- The `Set`, `List`, and `Map` interfaces allow adding and printing objects from collections, typically displaying them within square brackets (`[]`). To format output differently or perform computations while iterating through a collection, you can use the `Iterator` interface for forward-only iteration, especially with unordered collections like `Set`, or the `ListIterator` interface for bidirectional iteration in ordered collections like `List`.
- The Generics framework significantly improves the usability and safety of the Collections Framework in Java. It eliminates the need for casting when retrieving objects, ensures compile-time type safety, and allows for more concise and readable code. By declaring specific types in your collections, you can create homogeneous sets and avoid runtime errors related to type mismatches.
- The diamond operator (<>) introduced in Java SE7 simplifies the use of the Generics framework by allowing the compiler to infer the type of a collection from its declaration, eliminating the need to repeat the type on both sides of the assignment. This feature reduces redundancy and enhances code readability, as seen in examples like `Set<Person> mySet = new HashSet<>();`, where the type `Person` no longer needs to be specified twice.

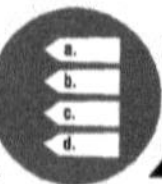

Multiple Choice Questions

1. Given the main method:

```
public static void main(String[] args)  {
  Set<Object> s = new HashSet<Object>();
  s.add("java");
  s.add(new String("java"));
  System.out.println(s);
}
```

Assume that the main method is in a valid Java source file and all the needed import statements are provided. Which of the following statements is TRUE?

a. The code will compile and run and will print "java" as the output.

b. The code will compile and run and will print "java" twice.

c. The code will not compile because adding duplicate objects in a Set is not allowed.

d. The code will compile, but will have a RuntimeException

2. What is the key difference between a `Set` and a `List`?

a. `List` maintains order, while `Set` ensures uniqueness

b. Both maintain element order

c. Both allow duplicate elements

d. `Set` allows duplicate elements, while `List` does not

3. **Which of the following statements IS NOT TRUE?**
   a. Objects in a Map can have duplicate keys.
   b. Objects in a set are all unique.
   c. Objects inside a list are ordered and can contain duplicates.
   d. You can add objects inside a Map as long as it is a key-value pair.

4. **In which package can we find the Collection Interface?**
   a. `java.io` package
   b. `java.util` package
   c. `java.awt` package
   d. `java.lang` package

5. **What characteristics of arrays make them different from collections?**
   a. Arrays can hold only objects
   b. Arrays are thread-safe
   c. Arrays can grow and shrink dynamically
   d. Arrays have a fixed size once created

6. **Which of the following statements about the `Set` interface is true?**
   a. A `Set` allows for duplicate elements but not null elements.
   b. A `Set` maintains the order of elements as they are inserted.
   c. A `Set` does not allow duplicate elements and permits at most one null element.
   d. A `Set` can contain duplicate keys but unique values.

7. **What will be the output of the following code snippet?**

```
Set<String> mySet = new HashSet<>();
mySet.add("apple");
mySet.add("banana");
mySet.add("apple");
System.out.println(mySet.size());
```

a. 1
b. 2
c. 3
d. Compilation error

8. **Consider the following code snippet. Which statement is true?**

```
Set<Integer> numbers = new HashSet<>();
    numbers.add(3);
    numbers.add(1);
    numbers.add(2);
    numbers.add(null);
    System.out.println(numbers);
```

a. The code will compile and run without errors, and the `HashSet` will contain `[null, 1, 2, 3]`.
b. The code will compile, but it will throw a `NullPointerException` at runtime.
c. The code will not compile due to a null being added to the `HashSet`.
d. The code will compile and run, but the `HashSet` will ignore the null value.

9. **Which of the following statements about the `List` interface is false?**
   a. A `List` allows duplicate elements.
   b. The `List` interface extends the Collection interface.
   c. `List` maintains the order in which elements are inserted.
   d. `List` provides a method to retrieve elements in reverse order without using any additional data structures.

10. **Given the code snippet below, what will be the output? Feel free to look at the API Documentation to see how to use the `ArrayList` methods.**

```
List<String> list = new ArrayList<>();
list.add("A");
list.add("B");
list.add("C");
list.set(1, "D");
list.add(1, "E");
System.out.println(list);
```

   a. `[A, B, D, C]`
   b. `[A, D, E, C]`
   c. `[A, E, D, C]`
   d. `[A, E, B, D, C]`

Multiple Choice Answer Key

1 – a	2 – a	3 – a	4 – b	5 – d
6 – c	7 – b	8 – a	9 – d	10 – c

Coding Task 1: Library Management System Using Java Collections

Task Overview:

Design and implement a basic library management system that can store, manage, and retrieve information about books. You will utilize Java Collections to handle the storage and manipulation of books. Each book in the system should have a unique identifier, a title, an author, and a genre.

Task Breakdown:

1. Creation of the `Book` Class:

   Create a Book class with the following attributes:

   - `String title`
   - `String author`
   - `String isbn` (each book's unique identifier)
   - `String genre`
   - Implement constructors, accessor methods, mutator methods, and a `toString()` method for easy display of book information.

2. Creation of the `Library` Class:

   Use a `Set<Book>` to maintain the collection of books, ensuring that no two books with the same ISBN can exist in the collection.

   Implement the following methods:

   - `void addBook(Book book)`: Adds a new book to the library.

- `Book searchByISBN(String isbn)`: Retrieves a book based on its ISBN. Returns `null` if the book is not found.
- `List<Book> searchByAuthor(String author)`: Returns a list of books by the specified author.
- `List<Book> searchByGenre(String genre)`: Returns a list of books that belong to the specified genre.
- `void removeBook(String isbn)`: Removes a book from the library using its ISBN.
- `void displayBooks()`: Displays all books in the library using an `Iterator`.

3. `Main` Class:

In the `main` method, showcase the use of the library system by:

- Adding several books to the library.
- Searching for a book by its ISBN.
- Finding books by a particular author or genre.
- Removing a book using its ISBN.
- Displaying all books in the library.

Additional Guidelines:

- Optimize operations for efficiency.
- Handle edge cases, such as attempting to add duplicate books or searching for books that do not exist.

Expected Results:

The program should successfully demonstrate the addition, retrieval, removal, and display of books within the library, using various Java Collections and Generics to manage the operations.

This task challenges students to apply their knowledge of Java Collections (`Set`, `List`, `Map`), Generics, and Iterators in a practical scenario involving the management of a collection of books.

Coding Task 2: Simple Login App

Create a Java Program that will read a text file named: `userList.txt`. The filename should be passed via the command line. It contains a list of usernames and passwords. The odd-numbered lines contain the usernames, while the even-numbered lines contain the passwords for the usernames in the previous lines. Here's a sample content for the list of `Strings` in the `userList.txt`:

```
user1
password1
user2
password2
user3
password3
```

Use HashMap to access the username and password.
Read the entry from the userList.txt

Username	*Password*
user1	password1
user2	password2
user3	password3

```
Sample Runtime 1:

java Login userList.txt <ENTER>
Enter username: user1
Enter password: password1
Access Granted

Sample Runtime 2:

java Login userList.txt <ENTER>
Enter username: someValue
Enter password: somePassword
Access Denied
```

CHAPTER 9

Building Graphical User Interfaces (GUI) in Java

Key Learning Objectives

- Introducing the `java.awt` package
- Things to consider while building your GUI application
- Learning about Java layout managers
- Understanding layout managers while creating your `SimpleCalculator` app

This chapter dives into building graphical user interfaces (GUIs) using the components and tools provided in `java.awt` package. You'll learn how to incorporate essential GUI elements such as buttons, text fields, labels, and containers to construct user-friendly applications. We'll cover the roles of panels, frames, and dialogs, as well as introduce Java's various layout managers, such as `FlowLayout`, `BorderLayout`, and `GridLayout`, which help arrange components in a visually appealing manner. As we explore customizing layout managers, you will apply these concepts by building and refining a simple calculator application.

9.1 Introducing the `java.awt` Package

Welcome to the Java AWT (Abstract Window Toolkit) section, an essential part of the Java API. The AWT package allows you to create graphical user interface (GUI) components like buttons, frames, text fields, text areas, and more. It's important to note that AWT was the original GUI toolkit included when Java 1.0 was first released, marking it as a first-generation toolkit.

As Java evolved, the second-generation user interface toolkit called Swing was introduced, providing more advanced features and flexibility. In addition to AWT and Swing, Java now also offers JavaFX, another powerful toolkit for creating rich user interfaces. However, for this discussion, we'll focus on AWT.

To begin with, let's revisit a fundamental concept: in Java, the `Object` class is the root of all class hierarchies. One of the subclasses of `Object` is the `Component` class, which represents anything visual that you can interact with in a user interface. Examples of components include buttons, labels, checkboxes, choice boxes, list boxes, and even containers.

Figure 9.1 The AWT Class Hierarchy

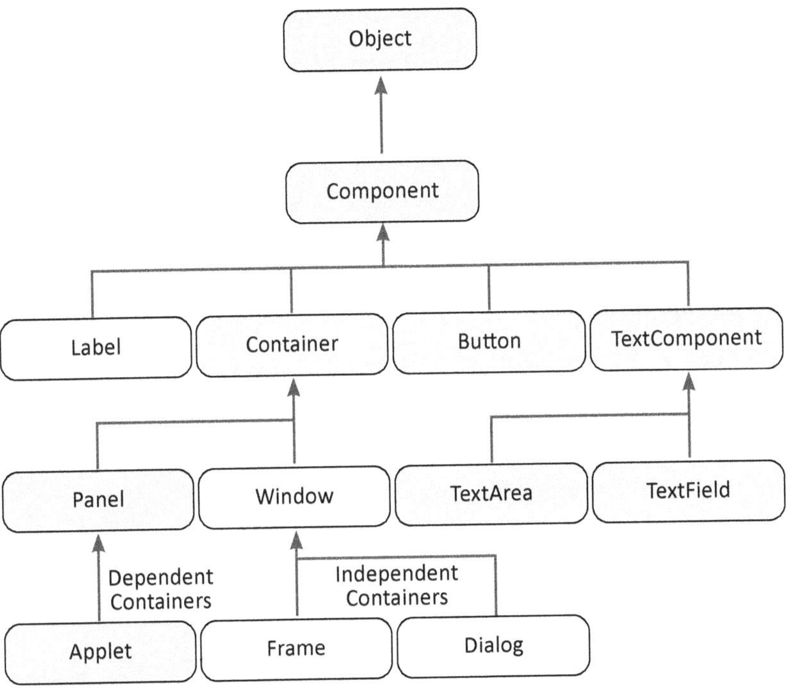

Figure 9.1 illustrates the class hierarchy of Java's Abstract Window Toolkit (AWT). At the top of the hierarchy is the `Object` class, from which all AWT components inherit. The `Component` class serves as the root for all graphical UI elements such as `Button`, `Label`, `Checkbox`, `Choice`, and `List`. These are individual user interface components.

The `Container` class, which also inherits from `Component`, can hold other components and containers, making it essential for creating complex UI layouts. `Container` branches into two main types:

- Window-based containers, such as `Frame` and `Dialog`, which are *independent* containers that do not need to be embedded in another UI component.

- Panel-based containers, such as `Panel` and `Applet`, which are *dependent* containers and must be added to a window or another container to be displayed.

This hierarchy shows how AWT organizes reusable and extendable UI components for building graphical user interfaces in Java applications.

Among these, the `Container` class is particularly important because it allows you to hold and organize other components within it. For example, you can place a button inside a container. Two common subclasses of the `Container` class are `Window` and `Panel`.

A `Window` is a type of container that can exist independently. Once you create a window, it can run on its own without needing any other containers. On the other hand, a `Panel` is also a container but cannot exist independently. It must be placed inside another container, such as a window or an applet, to function properly.

For instance, you can create a `Frame` (which is a type of window) and launch it by itself. However, if you create a `Panel`, it won't run on its own. Instead, it needs to be added to a frame or another container.

Each container in AWT comes with a default layout manager that controls how the components within it are arranged.

It's also worth noting that the `Applet` class, which was introduced in Java 1.1, is only supported up to Java 8. From Java 9 onwards, the `Applet` class has been deprecated, and as of Java 17, they have been completely removed.

Now, let's take a closer look at the AWT package. It includes several interfaces and classes that you'll frequently use in building GUIs. For example, the `Button` class allows you to create clickable buttons, while the `Checkbox` class lets you create selectable checkboxes. The `Frame` class, as mentioned earlier, is used to create window components that can stand alone. By understanding and utilizing these components and containers effectively, you'll be able to create functional and user-friendly graphical interfaces in Java.

9.2 Things to Consider While Building Your GUI Application

After exploring the `java.awt` package, which includes the essential classes for building graphical user interface (GUI) applications in Java, let us explore the three main categories: *components, containers,* and *layout managers.*

1. **Components** are the visual elements of your application, such as buttons, frames, dialog boxes, text areas, and text fields. These are the parts of your application that users will interact with directly.
2. **Containers** are a special type of component. Like other components, they are visual elements, but their main purpose is to hold and organize other components within them. Examples of containers include frames and panels.
3. **Layout managers** determine how the components are arranged within a container. They control the positioning and sizing of components inside containers, ensuring that your application's interface is organized and easy to use.

9.2.1 The `Frame` Class

Let's dive into the `Frame` class, which is a fundamental part of creating graphical user interfaces (GUIs) in Java. A frame is a top-level window that includes a border and a title bar, making it a container that can hold other GUI components.

By default, frames use a layout manager called `BorderLayout`. However, you can easily change this layout manager to suit your needs, and we'll explore how to do that shortly. Frames are resizable, meaning you can adjust their size using the corners of the window.

Now, let's look at a sample code that demonstrates how to create a simple GUI using a frame. To begin, look at the source code provided at the end of this discussion, saved as `SampleFrame.java`. In the provided source code, we must import the necessary package, which is `java.awt` located on line 1.

In the sample code, a `Frame` object is declared, and the constructor and methods are defined in lines 9 to 12. This approach is the best practice when building a GUI application. First, you declare all your components and containers. Then, within the constructor, you initialize them, just like in the provided code. Finally, you define methods to manage the behavior of these components.

In the `startApp()` method, which is a user-defined method, we set up the properties for the `frame`. For example, we set the background color to yellow using `setBackground(Color.YELLOW)`, and we specify the size of the frame as 200 pixels wide and 200 pixels high using `setSize(200, 200)`. You may check the API documentation to see the common `Color` objects, which you can see under the `java.awt.Color` class.

Table 9.1 shows the predefined colors available in Java's `java.awt.Color` class, along with its corresponding RGB values. These colors can be accessed using constants from the Color class, such as `Color.RED`, `Color.BLUE` or through the constructor call, ie, `new Color (0, 255, 0)` for `Green`.

Table 9.1 Common Color Names and Their RGB Values

Color	Red	Green	Blue	RGB Value
Black	0	0	0	(0, 0, 0)
Blue	0	0	255	(0, 0, 255)
Cyan	0	255	255	(0, 255, 255)
Dark Gray	64	64	64	(64, 64, 64)
Gray	128	128	128	(128, 128, 128)
Green	0	255	0	(0, 255, 0)
Light Gray	192	192	192	(192, 192, 192)
Magenta	255	0	255	(255, 0, 255)
Orange	255	200	0	(255, 200, 0)
Pink	255	175	175	(255, 175, 175)
Red	255	0	0	(255, 0, 0)
White	255	255	255	(255, 255, 255)
Yellow	255	255	0	(255, 255, 0)

By default, frames are invisible when first created. It's crucial to call the `setVisible(true)` method, which makes the frame visible. Without this, your GUI will run, but you won't see anything on the screen.

After running the code, you'll see how the frame looks on your screen. Remember, the appearance of your GUI might slightly differ depending on whether you are using Windows, macOS, or another operating system.

Here's a simple example code for creating a basic frame. The source code below is saved as `SampleFrame.java`.

```
import java.awt.*;

public class SampleFrame
{
  // Attributes
  private Frame f;

  // Constructor
  public SampleFrame()
  {
      f = new Frame("My First GUI App");
  }

   // Methods
   public void startApp()
   {
       f.setBackground(Color.YELLOW);
       f.setSize(200, 200);
       f.setVisible(true);
   }

   public static void main(String args[])
   {
       SampleFrame sf = new SampleFrame();
       sf.startApp();
   }
} // end of code
```

Here's the runtime output of `SampleFrame.java`:

Figure 9.2 Runtime Output of *`SampleFrame.java`*

Frames in Java are containers, but there are also other containers, such as panels. Panels provide a space where you can place other components, but unlike frames, they cannot be launched on their own. Panels must be added to another container, like a frame.

In our next example, we will create both a frame and a panel. We set the frame's size to 400 by 200 pixels and its background color to yellow. We also create a panel with a size of 100 by 100 pixels and a purple background. Panels are placed inside frames using the `add()` method.

The `setBackground()` method changes the background color of a container, using either predefined colors like `Color.YELLOW` or custom colors by specifying RGB values. The RGB values range from 0 to 255 for each of the red, green, and blue components.

If you want to disable the default layout manager (which is `BorderLayout` for frames), you can call `setLayout(null)` to disable the layout manager. Then, you can manually position your components inside the frame.

Here's a simple example code for creating a basic frame and a panel. The source code below is saved as `PanelInsideAFrame.java`.

```
import java.awt.*;

public class PanelInsideAFrame {
    private Frame f;
    private Panel p;

    public PanelInsideAFrame() {
        f = new Frame("Panel in a Frame");
        p = new Panel();
    }

    public void startApp() {
        f.setSize(400, 200);
        f.setBackground(Color.YELLOW);

        p.setSize(100, 100);
        p.setBackground(new Color(255, 0, 255));

        // frame won't be using any layout managers --> null
        f.setLayout(null);
        f.add(p);
        f.setVisible(true);
    }

    public static void main(String args[]) {
        PanelInsideAFrame pan = new PanelInsideAFrame();
        pan.startApp();
    }
} // end of code
```

Try running this code on your own, and experiment with changing the frame's size and background color. You can also adjust the panel's size and color, using values between 0 and 255 for the RGB components.

Figure 9.3 shows the runtime output of the file `PanelInsideAFrame.java`.

Figure 9.3 Runtime Output of `PanelInsideAFrame.java`

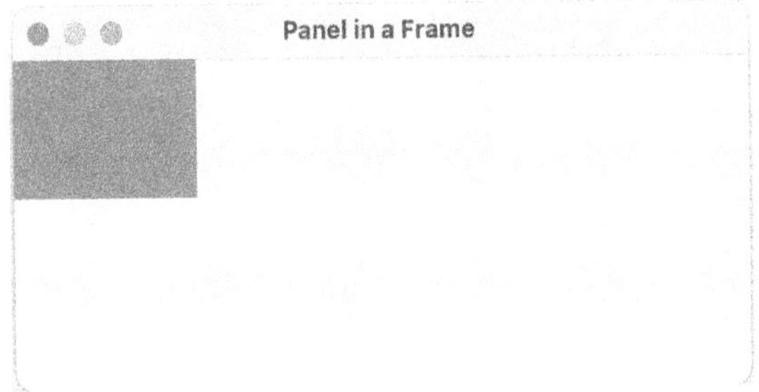

9.3 Java Layout Managers

Let's discuss layout managers, which are tools used in Java to arrange and position components within containers. These managers help control how your components are displayed on the screen.

For more complex layouts, you can nest containers, which means placing one container inside another, allowing you to create customized layouts. In the `java.awt` package, there are five main layout managers: `FlowLayout`, `BorderLayout`, `GridLayout`, `CardLayout`, and `GridBagLayout`. Additionally, the Swing package offers two more layout managers: `BoxLayout` and `SpringLayout`.

9.3.1 The `FlowLayout` Manager

Let's start by understanding how the `FlowLayout` manager works. `FlowLayout` arranges components in a row, much like words in a line of text. By default, it arranges buttons or similar components in a horizontal line until it runs out of space, at which point it moves to the next line. The alignment of these lines can be adjusted using properties like `LEFT`, `RIGHT`, `CENTER`, `LEADING`, and `TRAILING`.

Here's an example to illustrate this. Imagine you have four buttons. With `FlowLayout`, if only two buttons fit in the first row, the third button automatically moves to the next row, and so on. If you resize the window (the container), the buttons will adjust themselves to fit the new size.

Below is a sample code demonstrating the use of `FlowLayout`: The filename is `FlowLayoutExample.java`.

```
import java.awt.*;

public class FlowLayoutExample {

    // Attributes
    private Frame f;
    private Button addBtn, subBtn, mulBtn, divBtn;

    // Constructor
    public FlowLayoutExample() {
        f = new Frame("Flow Layout Example");
        addBtn = new Button("+");
        subBtn = new Button("-");
        mulBtn = new Button("*");
        divBtn = new Button("/");
    }
```

```

    // Methods
    public void startApp() {
        // setting FlowLayout as the layout manager
        f.setLayout(new FlowLayout());
        f.add(addBtn);
        f.add(subBtn);
        f.add(mulBtn);
        f.add(divBtn);
        f.setSize(200, 200);
        f.setVisible(true);
    }

    public static void main(String[] args) {
        FlowLayoutExample example
                        = new FlowLayoutExample();
        example.startApp();
    }
}
```

When we try to compile and run the code, we'll have this output below. Try resizing the frame horizontally and vertically, and observe how the buttons are realigned within the frame. The images below show the different versions of the `Frame` when resized.

Figure 9.4 FlowLayout Sample Output 1

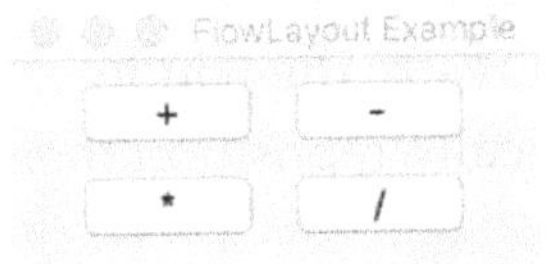

Figure 9.5 FlowLayout Sample Output 2

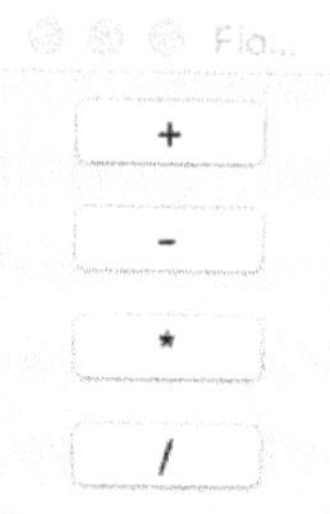

Figure 9.6 FlowLayout Sample Output 3

In this code, we create a frame and four buttons. Each button is labeled with a mathematical operator: addition, subtraction, multiplication, and division. After declaring and initializing the frame and buttons, we set the layout of the frame to `FlowLayout` using `f.setLayout(new FlowLayout());`. This line is crucial because if omitted, the frame will default to `BorderLayout`, which arranges components differently.

Once the layout manager is set, we add the buttons to the frame using the `add()` method. Finally, we set the size of the frame to 200x200 pixels and make it visible with `f.setVisible(true);`.

After compiling and running this code, you will see a window with four buttons arranged according to the `FlowLayout`. If you resize the window, you will notice how the buttons automatically reposition themselves to fit the available space.

`FlowLayout` also allows you to specify the alignment of the components within the layout. By default, the alignment is centered, but you can change it by passing a different alignment constant, like `FlowLayout.LEFT` or `FlowLayout.RIGHT` when setting the layout.

This is just one example of a layout manager provided by the `java.awt` package, and it gives you a glimpse of how powerful and flexible layout managers can be in organizing your GUI components.

9.3.2 The `BorderLayout` Manager

The next layout manager that we're going to explore is the default layout manager used in Java frames, known as the `BorderLayout`. This layout manager divides a container into five regions: `NORTH`, `SOUTH`, `EAST`, `WEST`, and `CENTER`. When arranging components in a `BorderLayout`, each region can hold only one component, and each region is specified by its corresponding constant (e.g., `BorderLayout.NORTH`, `BorderLayout.SOUTH`, etc.).

If you don't specify a region when adding a component, it will automatically be placed in the `CENTER` region. This region is the default and is the only one that can expand both horizontally and vertically. Components in the `NORTH` and `SOUTH` regions only resize horizontally, while those in the `EAST` and `WEST` regions only resize vertically.

Figure 9.7 BorderLayout Manager

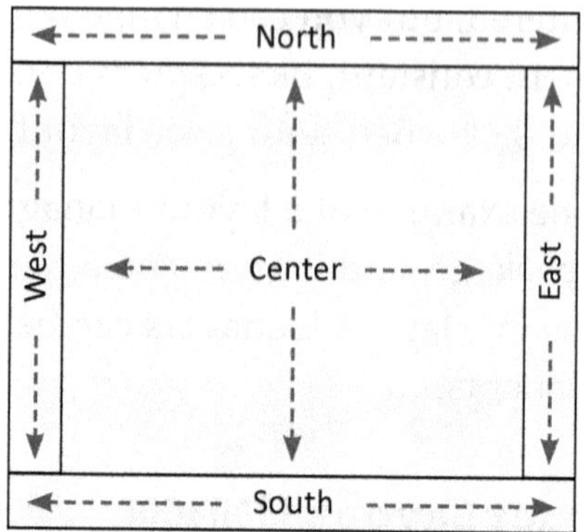

Let's go through an example. Suppose we have a frame with five buttons: button-north (`bNorth`), button-south (`bSouth`), button-center (`bCenter`), button-west (`bWest`), and button-east (`bEast`). In the constructor of our class, we would first create the frame and the buttons.

Here's a sample code with a filename: `BorderLayoutExample.java`.

```
import java.awt.*;

public class BorderLayoutExample
{
    private Frame f;
    private Button bNorth, bSouth, bCenter, bWest, bEast;

    public BorderLayoutExample()
    {
      f = new Frame("BorderLayout Example");

      bNorth = new Button("North");
      bSouth = new Button("South");
```

```
      bCenter = new Button("Center");
      bWest = new Button("West");
      bEast = new Button("East");
    }

    public void startApp()
    {
      // frames are using BorderLayout by default,
      // there's no need to change the layout manager
      // of the Frame f;

      f.add(bNorth, BorderLayout.NORTH);
      f.add(bSouth, BorderLayout.SOUTH);
      f.add(bCenter, BorderLayout.CENTER);
      f.add(bWest, BorderLayout.WEST);
      f.add(bEast, BorderLayout.EAST);

      f.setSize(400, 200);
      f.setVisible(true);
    }

    public static void main(String args[])
    {
       BorderLayoutExample be
                = new BorderLayoutExample();
       be.startApp();
    }
}
```

Next, we use a method, let's call it `startApp()`, to set up our frame. Since `BorderLayout` is the default layout manager for frames, we don't need to call the `setLayout()`

method as we did in the previous example. Instead, we can directly add the buttons to the frame, making sure to specify their positions.

If you forget to specify the region when adding a button, all buttons will stack on top of each other in the `CENTER` region, and only the last one added will be visible. To avoid this, always specify the correct region for each button.

When you run this code, you will see a frame with five buttons, each located in one of the five regions. If you resize the window horizontally, only the buttons in the `NORTH`, `CENTER`, and `SOUTH` regions will resize. If you resize the window vertically, only the buttons in the `WEST`, `CENTER`, and `EAST` regions will resize. The buttons in the `NORTH` and `SOUTH` regions will not change size when the window is resized vertically.

Figure 9.8 *`BorderLayoutExample.java`* **Runtime Output**

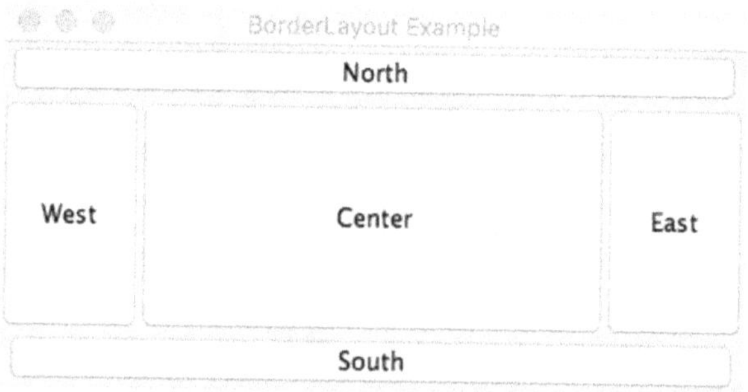

9.3.3 The `GridLayout` Manager

The `GridLayout` manager arranges components within a container in the form of a rectangular grid. The container is divided into equal-sized cells, with one component placed in each cell. When the container is resized, all the components within it remain equally sized, adjusting to fit the new size.

For example, let's say you have six buttons and you place them inside a frame that uses `GridLayout`. No matter how you resize the frame, all six buttons will adjust their size equally to fit within the frame. This makes `GridLayout` a great choice for building applications like a Minesweeper game or a calculator, where you need a consistent grid of buttons.

Now, let's look at a sample code saved as `GridLayoutExample.java`. In this example, we have a frame and six buttons. Inside the constructor, we create the frame and the six buttons.

```
import java.awt.*;

public class GridLayoutExample
{
    private Frame f;
    private Button b1, b2, b3, b4, b5, b6;

    public GridLayoutExample()
    {
        f = new Frame("GridLayout Example");

        b1 = new Button("B1");
        b2 = new Button("B2");
        b3 = new Button("B3");
        b4 = new Button("B4");
        b5 = new Button("B5");
        b6 = new Button("B6");
    }

    public void startApp()
    {
```

```
22          // to set the layout manager of our Frame f;
23          f.setLayout(new GridLayout(3, 2));
24
25          // put all the buttons inside the frame,
26          // left --> right
27          // top --> bottom
28
29          f.add(b1);
30          f.add(b2);
31          f.add(b3);
32          f.add(b4);
33          f.add(b5);
34          f.add(b6);
35
36          f.pack();
37          f.setVisible(true);
38      }
39
40      public static void main(String args[])
41      {
42          GridLayoutExample ge = new GridLayoutExample();
43          ge.startApp();
44      }
45  }
```

By default, frames use `BorderLayout`, so we need to change this by calling the `setLayout()` method on our frame and setting it to `new GridLayout()`. When creating a `GridLayout`, you can pass either two or four integer values to its constructor. In line 23 in the code, we used `new GridLayout(3, 2)`, which means that the layout manager will set up 3 rows and 2 columns.

If you don't pass any values in the constructor, the layout manager will create a single row with one component per column by default. However, in our code example, we passed two integer values to the constructor. These values specify the number of rows and columns for the grid. Alternatively, you can also specify the horizontal gap (`hGap`) and vertical gap (`vGap`) between components, which are measured in pixels.

In our example, we've set up the grid with 3 rows and 2 columns. When we add the buttons to the frame, they are arranged from left to right, top to bottom.

Instead of setting the size of the frame manually with `f.setSize()`, you can use a method called `f.pack()`. The `pack()` method automatically sizes the frame just large enough to fit all the components. If you have more components, the frame becomes larger; if you have fewer components, the frame will be smaller.

Finally, we set the frame to be visible by calling `setVisible(true)`. Now let's save and compile this code and run it.

Figure 9.9 `GridLayoutExample.java` **Runtime Output 1**

You can also resize the `Frame` with `GridLayout` as its layout manager. As you can see, each region is resized uniformly.

Figure 9.10 `GridLayoutExample.java` **Runtime Output 2**

Figure 9.11 `GridLayoutExample.java` **Runtime Output 3**

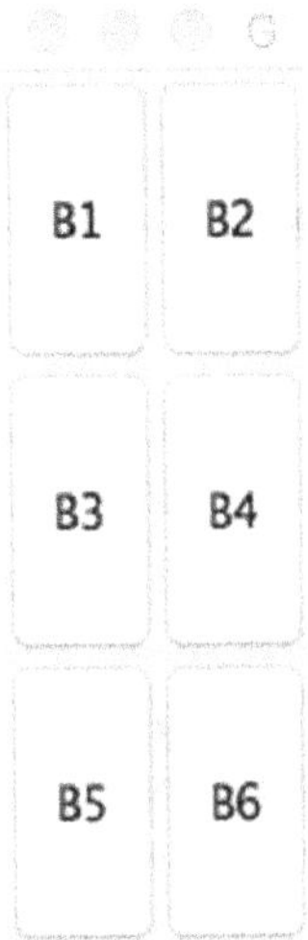

Here's what happens: The frame displays six equally sized buttons. If you resize the frame, you'll notice that all the buttons resize equally to fit the new dimensions.

But what happens if you have fewer buttons than the grid allows? For example, what if you only had five buttons instead of six? The frame will still display a 3x2 grid, but the last cell will be left empty.

The same happens if you only have four buttons. The grid layout will still maintain the 3x2 structure, with the extra spaces left blank.

If you try to add more than 6 buttons, the `GridLayout` will automatically add an extra column to accommodate the additional components. Therefore, the 7th button (and any additional ones) will be placed in a new column, altering the original layout of 3 rows and 2 columns.

When considering: `FlowLayout`, `BorderLayout`, and `GridLayout`, always remember that `FlowLayout` is the default layout manager for a `Panel`, while `BorderLayout` is the default for a `Frame`.

9.4 Understanding Layout Managers While Creating Your `SimpleCalculator` App

In this lesson, we'll learn how to customize our layout by combining these layout managers by nesting them. We want to create a `SimpleCalculator` that has a complex layout manager.

Figure 9.12 *`SimpleCalculator.java`* **Runtime Output**

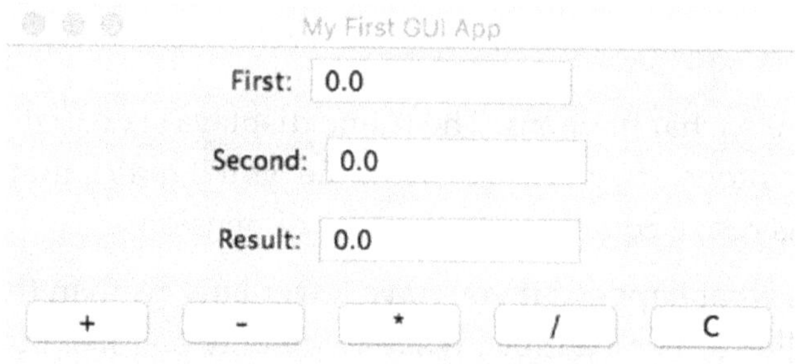

Choosing the Right Layout Manager

Before customizing your layout, check if an existing layout manager meets your needs. For example, `GridBagLayout` (from the AWT package) or `SpringLayout` and `BoxLayout` (from the Swing package) are very flexible. You might also find other layout managers online. To simplify layouts, you can use panels, which act as containers for grouping components.

Building a Simple Calculator

We'll start by creating a simple calculator as our first GUI application. This calculator will have:

- Three labels: `First:`, `Second:`, and `Result:`.
- Three text fields: We'll refer to them as `tf1`, `tf2`, and `tf3`.
- Five buttons: For adding, subtracting, multiplying, dividing, and clearing the input.

Setting up the application

1. Import Necessary Packages: Start by importing the AWT package:

```
import java.awt.*;
```

2. Define the class and variables: Next, create a class named `SimpleCalculator`, and define the necessary variables for the frame, labels, text fields, and buttons.

Figure 9.13 List of *Containers* and *Components* used for *SimpleCalculator.java*

```
Containers:
Frame     f;

Components:
Label     l1, l2, l3;
TextField tf1, tf2, tf3;
Button    bAdd, bSub, bMul, bDiv, bClear;
```

```
public class SimpleCalculator {

    private Frame f;

    private Label l1, l2, l3;

    private TextField tf1, tf2, tf3;
```

```
    private Button bAdd, bSub, bMul, bDiv, bClear;
    // Constructor
    public SimpleCalculator() {
        // Initialization goes here
    }
    // Method to start the application
    public void startApp() {
        // Code to start the app
    }
    public static void main(String[] args) {
        SimpleCalculator sc = new SimpleCalculator();
        sc.startApp();
    }
}
```

3. Initialize components: Inside the constructor, initialize the frame and components.

```
f = new Frame("My First GUI App");
l1 = new Label("First:");
l2 = new Label("Second:");
l3 = new Label("Result:");
tf1 = new TextField("0.0", 15);
tf2 = new TextField("0.0", 15);
tf3 = new TextField("0.0", 15);
bAdd = new Button("+");
bSub = new Button("-");
bMul = new Button("*");
bDiv = new Button("/");
bClear = new Button("C");
```

4. Use panels for layout: Create four panels to hold your components.

```
Panel p1 = new Panel();
Panel p2 = new Panel();
Panel p3 = new Panel();
Panel p4 = new Panel();
```

Figure 9.14 Assembling all the components together

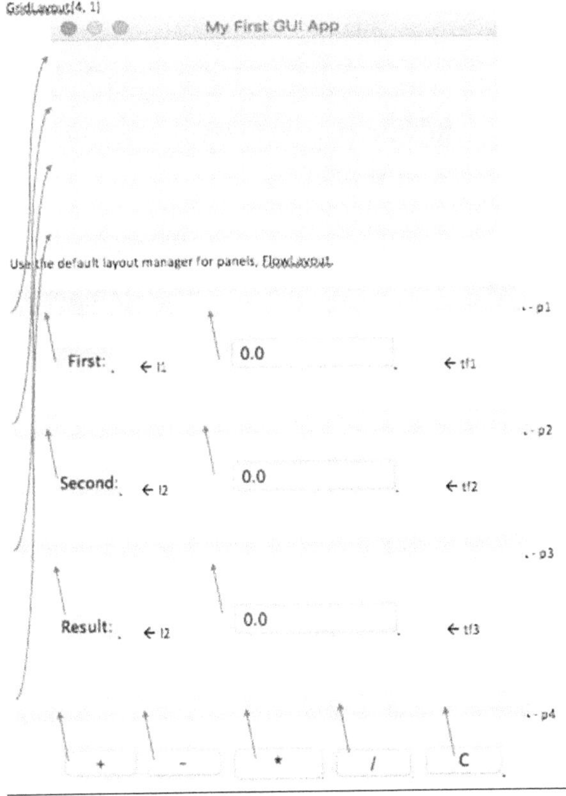

Add components to each panel.

```
p1.add(l1);
p1.add(tf1);
p2.add(l2);
p2.add(tf2);
p3.add(l3);
p3.add(tf3);
p4.add(bAdd);
p4.add(bSub);
p4.add(bMul);
p4.add(bDiv);
p4.add(bClear);
```

5. Set layout and add panels to frame: Set the layout for the frame and add the panels.

```
f.setLayout(new GridLayout(4, 1));
f.add(p1);
f.add(p2);
f.add(p3);
f.add(p4);
```

6. Display the frame: Finally, pack the frame and make it visible.

```
f.pack();
f.setVisible(true);
```

When you compile and run this code, you'll see a simple user interface with labels, text fields, and buttons arranged neatly. The frame automatically resizes to fit the components. However, note that this calculator currently only displays the UI; it doesn't perform any calculations or respond to button clicks yet. In the next chapter, we will discuss event handling techniques that will enable us to respond to button clicks to perform calculations.

Here's the final assembled code:

```
import java.awt.*;
public class SimpleCalculator
{
    // Containers
    private Frame f;
    private Panel p1, p2, p3, p4;

    // Components
    private Label l1, l2, l3;
    private TextField tf1, tf2, tf3;
    private Button bAdd, bSub, bMul, bDiv, bClear;

    public SimpleCalculator()
    {
        // container
        f = new Frame("My First GUI App");

        p1 = new Panel();
        p2 = new Panel();
        p3 = new Panel();
        p4 = new Panel();

        // components
```

```
        l1 = new Label("First: ");
        l2 = new Label("Second: ");
        l3 = new Label("Result: ");

        tf1 = new TextField("0.0", 15);
        tf2 = new TextField("0.0", 15);
        tf3 = new TextField("0.0", 15);

        bAdd = new Button("+");
        bSub = new Button("-");
        bMul = new Button("*");
        bDiv = new Button("/");
        bClear = new Button("C");
    }

    public void startApp()
    {
        // use the default layout manager for
        // all panels, use the FlowLayout.
        p1.add(l1);
        p1.add(tf1);

        p2.add(l2);
        p2.add(tf2);

        p3.add(l3);
        p3.add(tf3);

        p4.add(bAdd);
        p4.add(bSub);
        p4.add(bMul);
```

```
        p4.add(bDiv);
        p4.add(bClear);

        // change the layout manager for our Frame f;
        // use GridLayout(4, 1);
        f.setLayout(new GridLayout(4, 1));

        f.add(p1);
        f.add(p2);
        f.add(p3);
        f.add(p4);

        f.pack();
        f.setVisible(true);
    }

    public static void main(String args[])
    {
        SimpleCalculator sc = new SimpleCalculator();
        sc.startApp();
    }
}
```

When you compile and run this file, it should give you this output.

Figure 9.15 Runtime Output of `SimpleCalculator.java`

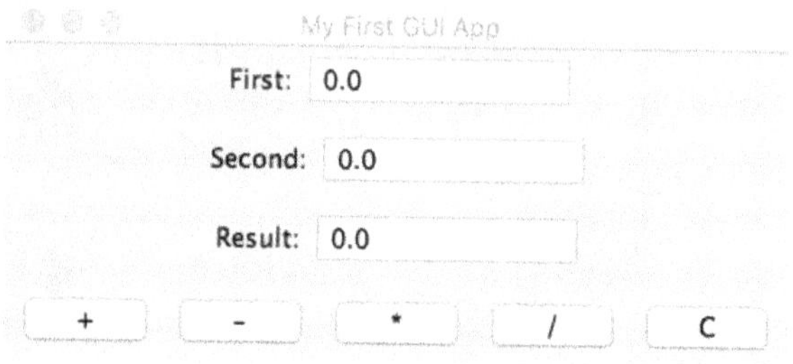

Chapter Summary

- Java's Abstract Window Toolkit (AWT) is the original API for creating graphical user interfaces, providing components like buttons, frames, and text fields. While newer toolkits like Swing and JavaFX offer more advanced features, this discussion focuses on AWT, where all visual elements are derived from the `Component` class, a subclass of `Object`.
- When building a Java GUI application, you'll work with components (like buttons and text fields), containers (which organize these components, such as frames and panels), and layout managers (which control their arrangement). Understanding how these elements interact is crucial for designing effective user interfaces.
- Java layout managers control the arrangement and size of components within a container, ensuring a consistent layout regardless of screen size or resolution. They provide various ways to organize GUI elements, such as arranging them in a grid, flowing them in a row, or positioning them in specific regions of a container. You can use any of the AWT layout managers (`FlowLayout`, `BorderLayout`, `GridLayout`, `CardLayout`, or `GridBagLayout`).
- Customizing layout managers in Java allows you to create tailored layouts by combining different managers or creating your own. This flexibility lets you design unique and complex interfaces that suit specific application needs beyond the standard layout options. This is usually achieved by nesting different layout managers together.

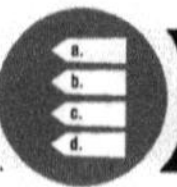

Multiple Choice Questions

1. **How do you change the current layout manager for a container?**
   a. Use the `setLayout()` method
   b. Once created, you cannot change the current layout manager of a component
   c. Use the `setLayoutManager()` method
   d. Use the `updateLayout()` method

2. **How do you indicate where a component will be positioned using Flowlayout?**
   a. Do nothing, the `FlowLayout` will position the component
   b. Pass an X/Y percentage parameter to the `add()` method
   c. Assign a row/column grid reference
   d. Assign a region: North, South, East, West

3. What best describes the appearance of an application with the following code?

```
import java.awt.*;
public class FlowAp extends Frame{
  public static void main(String args[]){
    FlowAp fa=new FlowAp();
    fa.setSize(400,300);
    fa.setVisible(true);
  }
  FlowAp(){
    add(new Button("One"));
    add(new Button("Two"));
    add(new Button("Three"));


    add(new Button("Four"));
  }//End of constructor
}//End of Application
```

a. A `Frame` with buttons marked One to Four placed on each edge.

b. A `Frame` with one large button marked Four in the Centre

c. A `Frame` with buttons marked One to four running from the top to bottom

d. An `Error` at runtime indicating you have not set a LayoutManager

4. What most closely matches the appearance when this code runs?

```
import java.awt.*;
public class ComplexLayout extends Frame{
  public static void main(String args[]){
     ComplexLayout cl = new ComplexLayout();
  }
  ComplexLayout(){
     Panel p = new Panel();
     p.setBackground(Color.pink);
     p.add(new Button("One"));
     p.add(new Button("Two"));
     p.add(new Button("Three"));
     add("South", p);
     setLayout(new FlowLayout());
     setSize(300,300);
     setVisible(true);
  }
}
```

a. The buttons will not be displayed

b. Only button three will show occupying all of the frame

c. The buttons will run from left to right along the bottom of the Frame

d. The buttons will run from left to right along the top of the Frame

5. What will be displayed when you attempt to compile and run the following code?

```
import java.awt.*;
public class MyButton extends Frame{
  public static void main(String argv[]){
     MyButton myButton = new MyButton();
  }
  MyButton(){
     Button helloButton = new Button("Hello");
     Button byeButton = new Button("Bye");
     add(helloButton);
     add(byeButton);
     setSize(300,300);
     setVisible(true);
  }
}
```

a. One button occupying the entire frame, saying Hello
b. Two buttons at the top of the frame, one saying Hello, the other saying Bye
c. One button occupying the entire frame, saying Bye
d. Two buttons side by side occupying all of the frame, Hello on the left and Bye on the right

6. Which of the following packages should you import if you want to use the GridLayout class?

a. `java.awt.event.*;`
b. `java.awt.*;`
c. `java.awt.events.*;`
d. `javax.swing.*;`

7. **Given the following code:**

```
import java.awt.*;
public class MyFrame extends Frame{
  public static void main(String argv[]){
    MyFrame m = new MyFrame();
    m.setSize(300,200);
    m.setVisible(true);
  }
}
```

How could you set the frame surface color to pink?

a. `s.setColor(PINK);`
b. `s.setBackground(Color.pink);`
c. `s.Background(pink);`
d. `s.color = Color.pink;`

8. **Which of the following layout managers IS NOT a part of the java.awt package?**

a. `FlowLayout`
b. `GridLayout`
c. `BoxLayout`
d. `CardLayout`

9. **Which of the following is a Container object?**
   a. `TextField`
   b. `Button`
   c. `TextArea`
   d. `Frame`

10. **Which of the following is NOT needed when you build your GUI App?**
    a. Components
    b. Containers
    c. Layout Managers
    d. File

Multiple Choice Answer Key

1 – a	2 – a	3 – b	4 – d	5 – c
6 – b	7 – b	8 – c	9 – d	10 – d

Coding Tasks

- Create a Java Application that will ask a user to enter four (4) integer numbers between 0 and 255. These numeric values represent the primary colors (R) Red, (G) Green, (B) Blue, and (A) Alpha -- the opacity of the color.
- These values should be entered in a text field.
- You will need to provide a "COMPUTE" Button that will compute the color based on the values entered in the four text fields.
- A panel on the lower half of the frame should be provided to show the computed color based on the values from the four text fields. Check the API for the constructor of `java.awt.Color` class that accepts four (4) values, RGBA (Red, Green, Blue, and Alpha) values.

Here's the prescribed output:

My Color Calculator

Red: 0

Green: 0

Blue: 0

Alpha: 0

Compute Clear

CHAPTER 10

Event Handling Techniques

Key Learning Objectives

- Understand the **basic principles of event handling** in Java GUI applications
- Explore the **Delegation Event Model** and its components: event sources and listeners
- Learn how to implement **listener interfaces** to respond to various user actions
- Discover how to simplify event handling using **adapter classes, inner classes,** and **anonymous classes**
- Use **lambda expressions** for more concise and readable event-handling code
- Apply your knowledge by **building a simple calculator** using Java event-handling techniques

This chapter dives into the core principles of *event-handling techniques in Java,* providing a solid foundation for managing user interactions in graphical applications. You'll begin by exploring the *Delegation Event Model,* where an event source delegates the handling of events to a listener—allowing for cleaner, modular code.

We will cover the *fundamentals of event handlers* and examine different ways to respond to events using *listener interfaces,* as well as simplified strategies through *adapter classes, inner classes,* and *anonymous classes.* These techniques help you control program behavior based on user actions such as button clicks, mouse movements, or key presses.

To modernize your approach, we will also introduce *lambda expressions,* which allow for more concise and readable event-handling logic. By the end of the chapter, you will apply what you've learned by *building a simple calculator,* giving you practical experience in designing interactive, event-driven Java applications.

10.1 Fundamentals Of Event Handlers

Let's dive into the basics of event handling in Java. ***Events*** are simply objects that describe actions or occurrences, such as clicking a button, moving a mouse, or pressing keyboard shortcuts like CTRL-S. These user interactions are called events because they initiate something within the interface. Each interaction is tied to an ***event source,*** which refers to the component involved in the interaction, such as a button, window, or icon.

Figure 10.1 Event Handling Process Diagram

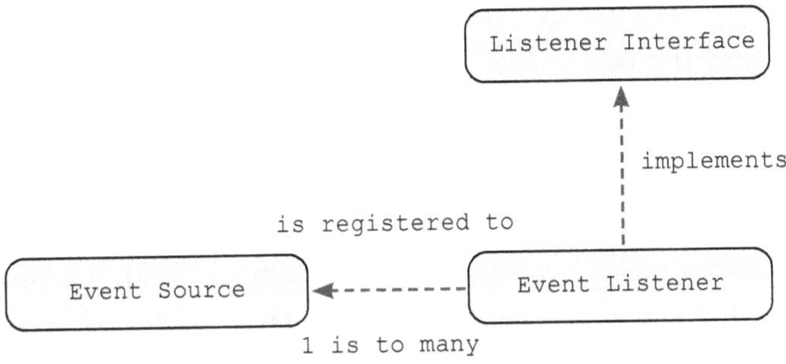

Figure 10.1 illustrates Java's Event Delegation Model, which is the foundation for handling events in graphical user interfaces. In this model, an *Event Source* is a component, such as a button, that generates events. An *Event Listener* is an object that wants to respond to those events. To do so, the listener must implement a specific *Listener Interface,* which defines the required event-handling methods. The listener is then *registered to the event source,* establishing a connection where the event source notifies the listener when an event occurs. This setup supports a *one-to-many relationship,* allowing a single event source to notify multiple listeners. By separating the event generation from its handling, the delegation model promotes a cleaner, more modular design.

For example, in a Word document, clicking the bold button or using the shortcut CTRL-B both trigger events. The components responsible for generating these events are known as event sources, and they notify listeners whenever an event occurs. Each event is encapsulated in an event object, and Java's `java.util.EventObject` class is the parent for all event objects, with subclasses like `ActionEvent` and `WindowEvent`.

Various components in your interface can produce different types of events. Buttons typically generate

`ActionEvent` objects, while Windows may send `WindowEvent` objects. To understand these interactions better, you can explore the `java.awt` package, where you'll find the necessary classes and interfaces to work with events. Different types of events, like `ActionEvent`, `AdjustmentEvent`, and `FocusEvent`, exist for various kinds of user interactions.

When an `ActionEvent` occurs, such as clicking a button or pressing the spacebar, the event is triggered, and an event handler responds. ***Event handlers*** are methods designed to manage the event and execute the intended behavior. So, for instance, clicking the bold button or pressing CTRL-B both trigger the same event handler to make text bold. These event handlers ensure that, regardless of how you interact with a component, the desired action is performed.

10.2 Event Handling Techniques

When a user interacts with a specific user interface component, known as the event source, an appropriate method, referred to as an event handler, handles the interaction. In the following sections, we will explore different techniques to create event handlers.

10.2.1 The Event Delegation Model

These techniques include the delegation model, which involves using listeners or interfaces, adapter or abstract classes, and defining inner classes within other classes to handle events. This approach allows you to delegate the event-handling logic to separate classes, resulting in a loosely coupled structure between the GUI components and the event handler logic. Each event handler class can reside in its own Java source file, promoting modularity and maintainability. You can create as many event handlers as needed; however, after defining them, it is essential to register them properly so they can respond to the events generated by the GUI components.

Figure 10.2 The Event Delegation Model

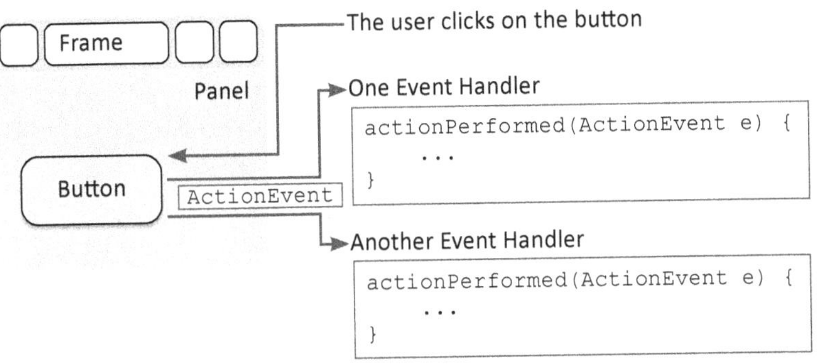

Figure 10.2 illustrates the Event Delegation Model in Java. When a user interacts with a component—such as clicking a button in a graphical user interface (GUI)—an event is generated, specifically an `ActionEvent` in this case. Instead of the button handling the event itself, the event is delegated to one or more external event handlers that implement the `actionPerformed(ActionEvent e)` method. These handlers contain the logic to respond to the user's action. This model promotes a clean separation between the GUI and the event-handling logic, allowing different classes or methods to process the same event in various ways depending on how they are registered.

This is the source code for your GUI, saved as: `MyGUI.java`

```
// MyGUI.java
import java.awt.*;
public class MyGUI
{
    private Frame f;
    private Button b;

    public MyGUI()
    {
        f = new Frame("Delegation Event Model");
        b = new Button("Click me!");
    }

    public void startApp()
    {
        f.add(b);
        f.pack();
        f.setVisible(true);

        // register your event handler

        // this is event registration #1
        b.addActionListener(new MyEvent());

        // this is event registration #2
        b.addActionListener(new MyEventToo());
    }

    public static void main(String args[])
    {
        MyGUI m = new MyGUI();
        m.startApp();
    }
}
```

Here's the source code for your event handler, saved as: `MyEvent.java`. This file was registered in `MyGUI.java` in line 24.

```
// MyEvent.java
import java.awt.event.*;

public class MyEvent implements ActionListener {
    public void actionPerformed(ActionEvent e)  {
    System.out.println("I was clicked!");
    }
}
```

Here's another source code for our second event handler, saved as: `MyEventToo.java`. This file was registered in `MyGUI.java` in line 27.

```
// MyEventToo.java
import java.awt.event.*;
public class MyEventToo implements ActionListener {
    public void actionPerformed(ActionEvent e)  {
        System.out.println("I was piggy backing!");
    }
}
```

You can create as many event handlers as you want, and they will all work as long as you register them with your GUI component.

10.2.2 The Use of Listeners

Listeners are interfaces. If you can recall the rules on interfaces, once you implement an interface, you are required to override ALL the methods in the interface. Table 10.1

shows the commonly used interfaces in the `java.awt.event` package.

Table 10.1 Listeners under the `java.awt.event` package

Category	Interface Name	Methods
Action	ActionListener	actionPerformed(ActionEvent)
Item	ItemListener	itemStateChange(ItemEvent)
Mouse	MouseListener	mousePressed(MouseEvent)
		mouseReleased(MouseEvent)
		mouseEntered(MouseEvent)
		mouseExited(MouseEvent)
		mouseClicked(MouseEvent)
Mouse motion	MouseMotionListener	mouseDragged(MouseEvent)
		mouseMoved(MouseEvent)
Key	KeyListener	keyPressed(KeyEvent)
		keyReleased(KeyEvent)
		keyTyped(KeyEvent)
Focus	FocusListener	focusGained(FocusEvent)
		focusLost(FocusEvent)
Adjustment	AdjustmentListener	adjustmentValueChanged (AdjustmentEvent)
Component	ComponentListener	componentMoved(ComponentEvent)
		componentHidden(ComponentEvent)
		componentResized(ComponentEvent)
		componentShown(ComponentEvent)
Window	WindowListener	windowClosing(WindowEvent)
		windowOpened(WindowEvent)
		windowIconified(WindowEvent)
		windowDeiconified(WindowEvent)
		windowClosed(WindowEvent)
		windowActivated(WindowEvent)

Category	Interface Name	Methods
		windowDeactivated(WindowEvent)
Container	ContainerListener	componentAdded(ContainerEvent)
		componentRemoved(ContainerEvent)
Text	TextListener	textValueChanged(TextEvent)

The complete set can always be seen in the API documentation. Here's a sample code that makes use of listeners. The filename used is: `MultiListeners.java.`

```
import java.awt.*;
import java.awt.event.*;

public class MultiListeners implements
      MouseListener, MouseMotionListener {
  private Frame f;
  private TextField tf;
  private Label label;

  public MultiListeners() {
    f = new Frame("I have Two listeners example");
    tf = new TextField(30);
    label = new Label("Click and drag the mouse");
  }

  public void startApp() {
    f.add(label, BorderLayout.NORTH);
    f.add(tf, BorderLayout.SOUTH);

    f.setSize(300, 200);
    f.setVisible(true);

    // event handler registration
```

```
      f.addMouseMotionListener(this);
      f.addMouseListener(this);
  }

      // The event handlers
      public void mouseDragged(MouseEvent e) {
        String s = "Mouse dragging:  X = "
          + e.getX() + " Y = " + e.getY();
        tf.setText(s);
      }

      public void mouseEntered(MouseEvent e) {
        String s = "Mighty Mouse is coming in..";
        tf.setText(s);
      }

      public void mouseExited(MouseEvent e) {
        String s = "The mouse has left the building";
        tf.setText(s);
      }
      // Unused event handlers
      // that needs to be overridden
      public void mouseMoved(MouseEvent e)
      {}
      public void mousePressed(MouseEvent e)
      {}
      public void mouseClicked(MouseEvent e)
      {}
      public void mouseReleased(MouseEvent e)
      {}

      public static void main(String args[]) {
```

```
57            MultiListeners ml = new MultiListeners();
58            ml.startApp();
59        }
60    }
```

As you can see in the implementation declaration in line 5, this sample code implements both `MouseListener` and `MouseMotionListener` interfaces. And since these interfaces are implemented, we have to override ALL the methods on both interfaces. Even if the methods are unused, they still need to be overridden.

10.2.3 The Use of Adapter Classes

There are situations when it is better to use adapter classes rather than listeners. Not all listeners have a corresponding adapter class. Only listeners that contain two or more methods have a corresponding adapter class. For example, if you implement `WindowListener`, you are required to override all seven methods even if you only need one. If this is the case, it is better to use the `WindowAdapter` instead.

Here's a sample code on how to use an adapter class with a file named `SampleAdapter.java`:

```
import java.awt.*;
import java.awt.event.*;

public class SampleAdapter extends WindowAdapter
{
    private Frame f;
    private TextField tf;
    private Label label;

```

```
    public SampleAdapter()
    {
        f = new Frame("A sample adapter code");
        tf = new TextField(30);
        label = new Label("You can quit this app");
    }

    public void startApp()
    {
        f.add(label, BorderLayout.NORTH);
        f.add(tf, BorderLayout.SOUTH);

        f.setSize(300, 200);
        f.setVisible(true);

        f.addWindowListener(this);
    }

    public void windowClosing(WindowEvent we)
    {
        System.exit(0);
    }

    public static void main(String args[])
    {
        SampleAdapter sa = new SampleAdapter();
        sa.startApp();
    }
}
```

Here's the output of the code `SampleAdapter.java`:

Figure 10.3 Screenshot of the runtime of the code *`SampleAdapter.java`*

A sample adapter code

You can quit this app

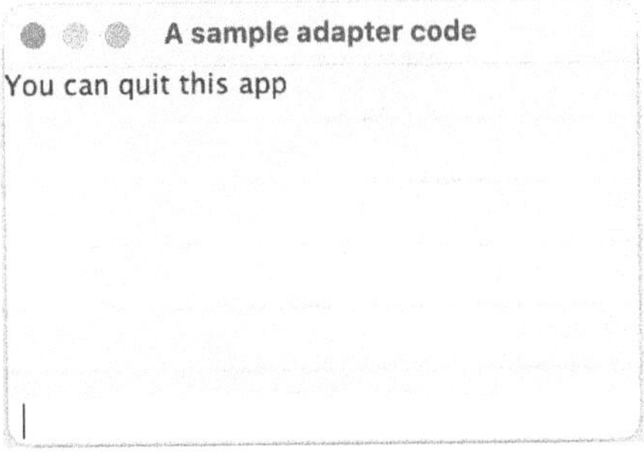

Since an adapter class is an abstract class, we have to use the keyword `extends`, which sometimes makes it a disadvantage to use an adapter class. Lines 28 to 31 are the only lines needed because you only need to override one method in the `WindowAdapter` class.

Table 10.2 shows a list of adapter classes in the `java.awt.event` package.

Category	Adapter Class	Methods
Mouse	MouseAdapter	mousePressed(MouseEvent)
		mouseReleased(MouseEvent)
		mouseEntered(MouseEvent)
		mouseExited(MouseEvent)
		mouseClicked(MouseEvent)
Mouse motion	MouseMotionAdapter	mouseDragged(MouseEvent)
		mouseMoved(MouseEvent)
Key	KeyAdapter	keyPressed(KeyEvent)
		keyReleased(KeyEvent)

Category	Adapter Class	Methods
		`keyTyped(KeyEvent)`
Focus	`FocusAdapter`	`focusGained(FocusEvent)`
		`focusLost(FocusEvent)`
Component	`ComponentAdapter`	`componentMoved(ComponentEvent)`
		`componentHidden(ComponentEvent)`
		`componentResized(ComponentEvent)`
		`componentShown(ComponentEvent)`
Window	`WindowAdapter`	`windowClosing(WindowEvent)`
		`windowOpened(WindowEvent)`
		`windowIconified(WindowEvent)`
		`windowDeiconified(WindowEvent)`
		`windowClosed(WindowEvent)`
		`windowActivated(WindowEvent)`
		`windowDeactivated(WindowEvent)`
Container	`ContainerAdapter`	`componentAdded(ContainerEvent)`
		`componentRemoved(ContainerEvent)`

10.2.4 Inner Classes

The purpose of inner classes is to make tightly coupled components. As for GUI apps, it is used to group the UI code and event handler code together in a single code base.

In an AWT GUI application, an inner class is often used for event handling because it allows the event listener to access the outer class's methods and fields without requiring additional references. This makes the code more organized and easier to manage. By defining an event handler as an inner class, the event-handling logic remains encapsulated within the main class, reducing clutter and improving readability. Additionally, since an inner class can access private members of its enclosing class, it simplifies interactions with UI components. For example, if a button click needs to update a label or change another component's state, an inner class can do this directly without requiring extra methods or references. This approach keeps the event-handling code close to the components it manages, making the application more structured and maintainable.

Using inner classes for event handling in AWT apps can make the code cleaner and more organized. An inner class can directly access the variables and methods of the outer class, which means you don't need to pass additional parameters to work with components. This reduces the amount of boilerplate code and makes the event-handling logic easier to understand and maintain. Additionally, because the inner class is tied to the outer class, it can be used specifically for handling events of that particular instance, keeping related code together.

Here's how we can implement inner classes in building your GUI app.

```

public class SimpleCalculator implements ActionListener
{
    ...    ...
    private class MyCloseButtonHandler extends WindowAdapter
    {
        public void windowClosing(WindowEvent we)
        {
            System.exit(0);
        }
    }
    ...    ...
}
```

10.2.5 Anonymous Classes

Anonymous classes allow you to create objects on the fly without giving them a formal name. These objects act as event handlers that do not require a separate class reference. Once the event handler has been executed, the anonymous object is automatically discarded, helping to conserve memory and improve performance.

This approach is particularly useful in building efficient mobile or desktop applications, where minimizing memory usage is key to ensuring smoother runtime performance. By reducing the number of active objects, the system can run more efficiently.

The general syntax for using an anonymous class is as follows:

```
// Syntax for anonymous class
ClassOrInterface obj = new ClassOrInterface() {
    // Override methods or add new code here
};
```

Here's a sample of how to implement anonymous classes in your GUI app.

```
// SampleAnonymousClassImplementation.java
import java.awt.*;
import java.awt.event.*;

public class SampleAnonymousClassImplementation {
    private Frame f;
    private TextField tf;
    private Label label;

```

```
10      public SampleAnonymousClassImplementation() {
11          f = new Frame("A sample anonymous code");
12          tf = new TextField(30);
13          label = new Label("You can quit this app");
14      }
15
16      public void startApp() {
17          f.add(label, BorderLayout.NORTH);
18          f.add(tf, BorderLayout.SOUTH);
19
20          f.setSize(300, 200);
21          f.setVisible(true);
22
23          // Using an anonymous class for event handling
24          f.addWindowListener(new WindowAdapter() {
25              public void windowClosing(WindowEvent we) {
26                  System.exit(0);
27              }
28          });
29      }
30
31      public static void main(String args[]) {
32          SampleAnonymousClassImplementation saci =
33              new SampleAnonymousClassImplementation();
34          saci.startApp();
35      }
36  }
```

As you can see in the code `SampleAnonymousClassImplementation.java`, we used an anonymous class declaration in lines 23 to 28.

10.2.6 The Use of Lambda Expression

Lambda expressions allow you to pass functions without names as arguments to methods. This provides a cleaner, more concise way to handle events compared to anonymous classes, making event handling simpler and more readable.

The general syntax for the use of the lambda expression is as follows:

```
component.addEventListener(event -> {
    // Code to handle the event
});
```

Here are two examples, with a single statement:

```
b.addActionListener(
    e -> System.out.println("Handled by "
                + "Lambda Expression"));
```

Here's with multiple statements:

```
b.addActionListener(
   e -> {
     System.out.println("My Event Handler");
     System.out.println("Handled by "
                 + "Lambda Expression");
   }
);
```

10.3 Building a `SimpleCalculator` App With Event Handlers

After we've built the structure of the **`SimpleCalculator`** app in the previous chapter using components from the `java.awt` package. The calculator consists of labels, text fields for input, and buttons for arithmetic operations. However, at this stage, the calculator is only a static interface—it doesn't perform any calculations yet. We are now ready to complete it by including the event handlers.

Our calculator's functionality is housed inside a constructor, where all the components are created. A method called `startApp()` arranges these components within the frame using a layout manager. Currently, the calculator looks functional, but doesn't respond to any user interactions. Clicking on any button (such as `+, -, *, /`) won't produce results, and the close button won't work either. Why? Because we haven't implemented any event handling yet.

Implementing Event Handling

To make the buttons functional, we need to implement event handling. Specifically, we'll use an `ActionListener` to capture button clicks. The `ActionListener` interface is part of the `java.awt.event` package, and it requires us to override the `actionPerformed()` method, which handles the action events.

Let's add this to our code before the `main()` method. The `actionPerformed()` method will determine which button was clicked by calling the `getSource()` method. Once we know the source of the event (the button), we can perform the appropriate action, whether it's adding, subtracting, multiplying, or dividing the numbers entered in the text fields.

Here's the actual code:

```
// SimpleCalculator.java
import java.awt.*;
import java.awt.event.*;

public class SimpleCalculator implements ActionListener {
    // Containers
    private Frame f;
    private Panel p1, p2, p3, p4;

    // Components
    private Label l1, l2, l3;
    private TextField tf1, tf2, tf3;
    private Button bAdd, bSub, bMul, bDiv, bClear;

    public SimpleCalculator() {
        // Frame and Panels
        f = new Frame("Simple Calculator");

        p1 = new Panel();
        p2 = new Panel();
        p3 = new Panel();
        p4 = new Panel();

        // Labels and Text Fields
        l1 = new Label("First Number: ");
        l2 = new Label("Second Number: ");
        l3 = new Label("Result: ");

        tf1 = new TextField("0.0", 15);
        tf2 = new TextField("0.0", 15);
```

```
        tf3 = new TextField("0.0", 15);

        // Buttons
        bAdd = new Button("+");
        bSub = new Button("-");
        bMul = new Button("*");
        bDiv = new Button("/");
        bClear = new Button("C");
    }

    public void startApp() {
        // Layout setup
        p1.add(l1);
        p1.add(tf1);
        p2.add(l2);
        p2.add(tf2);
        p3.add(l3);
        p3.add(tf3);
        p4.add(bAdd);
        p4.add(bSub);
        p4.add(bMul);
        p4.add(bDiv);
        p4.add(bClear);

        f.setLayout(new GridLayout(4, 1));
        f.add(p1);
        f.add(p2);
        f.add(p3);
        f.add(p4);
        f.pack();
        f.setVisible(true);

```

```
        // Register event listeners for buttons
        bAdd.addActionListener(this);
        bSub.addActionListener(this);
        bMul.addActionListener(this);
        bDiv.addActionListener(this);
        bClear.addActionListener(this);
       f.addWindowListener(new MyCloseButtonHandler());
    }

    // Event handler for button clicks
    public void actionPerformed(ActionEvent e) {
        Object source = e.getSource();
        double num1 = 0.0, num2 = 0.0, num3 = 0.0;

        if (!tf1.getText().isEmpty() &&
            tf2.getText().isEmpty()) {
            num1 = Double.parseDouble(tf1.getText());
            num2 = Double.parseDouble(tf2.getText());

            if (source == bAdd) {
                num3 = num1 + num2;
            } else if (source == bSub) {
                num3 = num1 - num2;
            } else if (source == bMul) {
                num3 = num1 * num2;
            } else if (source == bDiv) {
                num3 = num1 / num2;
            } else if (source == bClear) {
                tf1.setText("0.0");
                tf2.setText("0.0");
                tf3.setText("0.0");
            }
```

```
                tf3.setText(String.valueOf(num3));
            }
        }

        // Window close handler
        private class MyCloseButtonHandler
                        extends WindowAdapter {
            public void windowClosing(WindowEvent we) {
                System.exit(0);
            }
        }
        public static void main(String[] args) {
            SimpleCalculator sc = new SimpleCalculator();
            sc.startApp();
        }
}
```

This should give you a runtime output with working events. Figure 10.4 shows a sample screenshot of our `SimpleCalculator` App.

Figure 10.4 *`SimpleCalculator`* **App**

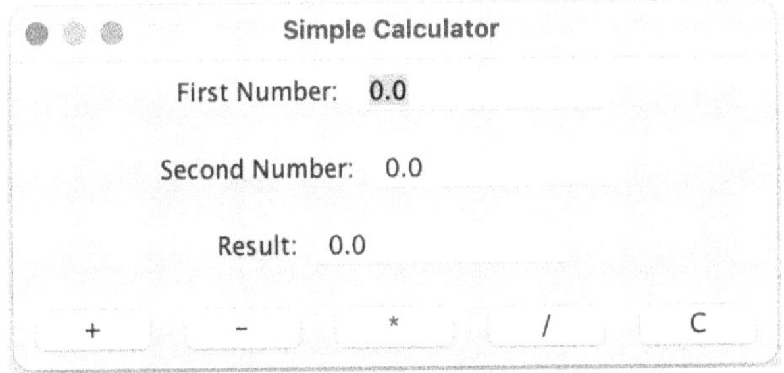

Chapter Summary

- Event handling in Java involves interactions such as button clicks, mouse movements, or keyboard shortcuts, which are captured as events and tied to specific components called event sources. Each event triggers a method, known as an event handler, that executes the desired action, ensuring consistent behavior across different types of user interactions.
- There are different event handling techniques, namely the delegation model, using the listener interface, using adapter classes, inner classes, anonymous classes, and using the lambda expression.

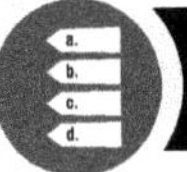

Multiple Choice Questions

1. What will happen when you attempt to compile and run this code?

```
import java.awt.*;
import java.awt.event.*;
public class MyWindowClosing extends
                Frame implements WindowListener{
   public static void main(String args[]){
      MyWindowClosing mwc = new MyWindowClosing();
   }
   public void windowClosing(WindowEvent we){
      System.exit(0);
   }//End of windowClosing
   public void MyWindowClosing(){
      setSize(300,300);
      setVisible(true);
   }
}//End of class
```

a. Error at compile time
b. Visible Frame created that can be closed
c. Compilation but no output at runtime
d. Error at compile time because of a comment before import statements

2. If event handling classes need to be used in your program, you need to import what Java package?

a. `javax.awt.events` package
b. `java.awt` package
c. `javax.events` package
d. `java.awt.event` package

3. **Which of the following DOES NOT generate an event object?**
   a. Clicking a button
   b. Inspecting the contents of a Frame object
   c. Typing in a text area
   d. Pressing the Enter key in a text field

4. **Which of the following is an abstract class found in the `java.awt.event` package?**
   a. `ActionListener`
   b. `MouseMotionListener`
   c. `MouseAdapter`
   d. `MouseListener`

5. **What do you call the object that refers to the interaction between the user and the GUI component?**
   a. Adapter Classes
   b. Event Listeners
   c. Event Handlers
   d. Events

6. **Which of the following is TRUE about event handlers?**
   a. Event handlers represent the interaction between the user and the GUI components.
   b. Event handlers are the methods invoked when an event is triggered.
   c. Event handlers are objects that send out event objects to all registered listeners when that event occurs.
   d. Event handlers are ALWAYS within the same code where your GUI components are.

7. **Which of the following is the best event handling strategy for event handlers if you want to reuse your event handlers for multiple GUI components coded in different Java source codes?**
   a. Extending an adapter class
   b. Using an anonymous class for your event handler
   c. Delegating your event handler to a different Java source code
   d. Using an inner class for your event handler

8. **What keyword should we use for the code snippet below?**

```
private class MyEventHandler ________ WindowAdapter
{
    // some codes here
}
```

   a. `extends`
   b. `implements`
   c. overload
   d. override

9. **Which of the following operators should you use when using the lambda expression as your event handling technique?**
   a. `->`
   b. `=>`
   c. `-=>`
   d. `==>`

10. Which of the following IS NOT a listener interface?

a. `MouseListener`
b. `MouseAdapter`
c. `MouseMotionListener`
d. `ActionListener`

Multiple Choice Answer Key

1 – a	2 – d	3 – b	4 – c	5 – d
6 – b	7 – c	8 – a	9 – a	10 – b

Coding Tasks

Coding Task 1: My Color Calculator (with Event Handler)

- Revisit the Coding Task in Chapter 9. Provide the needed functionality on your Color Calculator. It should be able to accept the (Red, Green, Blue, Alpha) values from the user's input on their respective textfields. The Red, Green, and Blue values are the primary colors, while Alpha represents the transparency level of a color. It controls how opaque or transparent a color should be. The red, green, blue, and alpha value ranges from 0 to 255. If the alpha value is 0, it means it is fully transparent (completely see-through), and 255 means fully opaque (no transparency).
- Catch invalid values entered in the text fields. Throw an exception if the user entered any of the following:
  - A negative value
  - A value beyond the range from 0 to 255
  - A non-integer value
  - A blank text field.
- Update the background color of the `Panel` with the computed values from the four text fields. You may want to check out the constructor from the `Color` class in the API where you can pass four integer values to instantiate the color.

Here's a sample output:

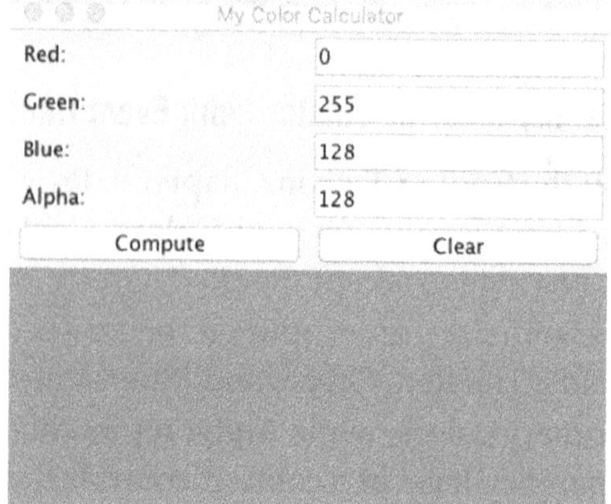

CHAPTER 11

Building a Swing Desktop App

Key Learning Objectives

- Understand the key differences between AWT and Swing in Java GUI development
- Learn how to convert AWT applications into Swing-based applications
- Explore commonly used Swing components, such as `JCheckBox`, `JRadioButton`, `JComboBox`, `JList`, and `JLabel`
- Gain hands-on experience in designing graphical user interfaces using Swing
- Learn how to package and distribute your Swing applications as Java Archive (`JAR`) files

In our final chapter, we explore the key differences between ***AWT and Swing***, focusing on how Swing offers a more flexible and modern approach to building graphical user interfaces in Java. While AWT laid the foundation for GUI development, Swing enhances it with a richer set of components and greater control over look and feel. You'll learn how to transition an existing AWT-based application into a Swing-based one, unlocking more advanced features and improved visual design. We'll also take a closer look at several commonly used Swing components—such as `JCheckBox`, `JRadioButton`, `JComboBox`, `JList`, and `JLabel`—and see how they can enhance user interaction. Finally, the chapter will guide you through packaging your Swing application into a Java Archive (JAR) file, making it easy to share, run, or deploy across platforms. By the end of this chapter, you'll be ready to design, build, and distribute full-featured Java desktop applications using Swing.

11.1 AWT vs. Swing

Java AWT (Abstract Window Toolkit) and Swing are both used for building graphical user interfaces (GUIs) in Java, but they have key differences.

Swing is part of the Java Foundation Classes (JFC) and is built on top of AWT. While AWT relies on native operating system components, Swing provides lightweight, platform-independent components that look the same across all operating systems.

In AWT, components like `Frame` and `Button` are directly tied to the platform, meaning their appearance can vary depending on whether you're using Windows, macOS, or Linux. In contrast, Swing components like `JFrame` and `JButton` look consistent across different platforms since they don't depend on the native OS.

Figure 11.1 `AWT` **Sample Frame for both Solaris and Windows OS**

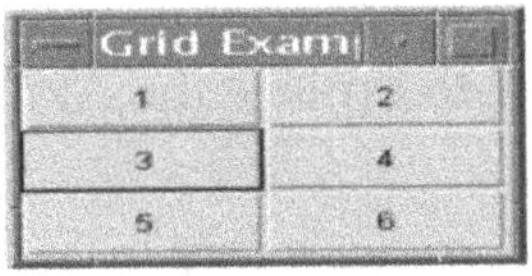

Solaris OS

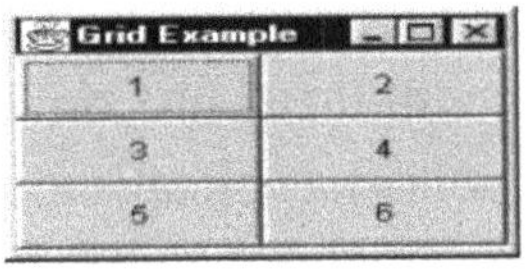

Microsoft Windows

Figure 11.2 `Swing` **Sample Frame**

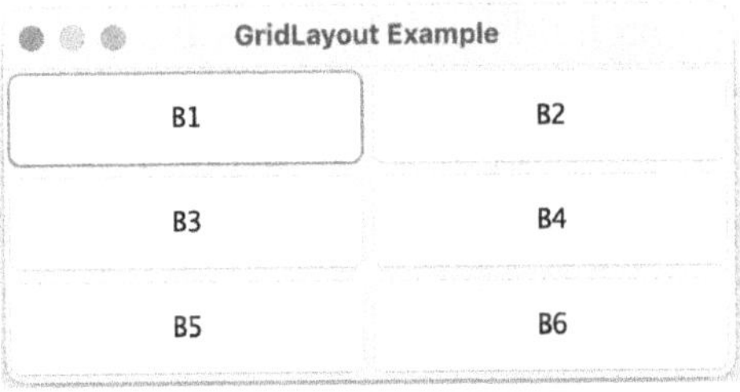

Swing adds more features and flexibility. It offers additional components like `JRadioButton`, `JComboBox`, `JCheckBox`, and `JLabel`, which are not available in AWT. Swing also supports a pluggable look-and-feel, meaning you can customize the appearance of components, something AWT does not support.

Another key difference is that AWT components are generally heavier in memory and slower, while Swing components are lightweight, offering better performance. Swing also follows the *MVC (Model-View-Controller)* design pattern, which AWT doesn't support.

In summary, AWT provides basic, platform-dependent components, while Swing offers a richer, more flexible, and platform-independent set of components. Swing was also introduced in Java 1.2 to replace the AWT package.

11.2 Converting Your AWT App to a Swing-Based App

In this section, we will focus on converting an AWT (Abstract Window Toolkit) application into a basic Swing-based application. To achieve this, it is essential to import the `javax.swing` package into our simple calculator application.

As a brief recap, the original calculator application utilizes various AWT components, including panels, frames, labels, text fields, and buttons. It also incorporates the AWT event package to manage event handling, such as using an `ActionListener` to respond to button clicks or implementing a close button handler that enables the application to close when the user presses `Command+Q` (for macOS) or `Alt+F4` (for Windows). This forms a fully functional AWT application.

Here's the original AWT `SimpleCalculator.java` source code:

```
// SimpleCalculator.java
import java.awt.*;
import java.awt.event.*;

public class SimpleCalculator implements ActionListener
{
    // Containers
    private Frame f;
```

```
    private Panel p1, p2, p3, p4;

    // Components
    private Label l1, l2, l3;
    private TextField tf1, tf2, tf3;
    private Button bAdd, bSub, bMul, bDiv, bClear;

    public SimpleCalculator()
    {
        // container
        f = new Frame("My First GUI App");

        p1 = new Panel();
        p2 = new Panel();
        p3 = new Panel();
        p4 = new Panel();

        // components
        l1 = new Label("First: ");
        l2 = new Label("Second: ");
        l3 = new Label("Result: ");

        tf1 = new TextField("0.0", 15);
        tf2 = new TextField("0.0", 15);
        tf3 = new TextField("0.0", 15);

        bAdd = new Button("+");
        bSub = new Button("-");
        bMul = new Button("*");
        bDiv = new Button("/");
        bClear = new Button("C");
    }

```

```
public void startApp()
{
    // use the default layout manager
    // for all panels, use the FlowLayout.

    p1.add(l1);
    p1.add(tf1);

    p2.add(l2);
    p2.add(tf2);

    p3.add(l3);
    p3.add(tf3);

    p4.add(bAdd);
    p4.add(bSub);
    p4.add(bMul);
    p4.add(bDiv);
    p4.add(bClear);

    // change the layout manager for our
    // Frame f; use GridLayout(4, 1);
    f.setLayout(new GridLayout(4, 1));

    f.add(p1);
    f.add(p2);
    f.add(p3);
    f.add(p4);

    f.pack();
    f.setVisible(true);

    // register your event handler to
```

```
        // your UI (event source)
        bAdd.addActionListener(this);
        bSub.addActionListener(this);
        bMul.addActionListener(this);
        bDiv.addActionListener(this);
        bClear.addActionListener(this);

        f.addWindowListener(new MyCloseButtonHandler());
    }

    // event handler
    public void actionPerformed(ActionEvent e)
    {
        Object source = e.getSource();
        double num1 = 0.0, num2 = 0.0, num3 = 0.0;

        if (tf1.getText() != null
                && tf2.getText() != null) {
          num1 = Double.parseDouble(tf1.getText());
          num2 = Double.parseDouble(tf2.getText());

          if (source == bAdd)
          {
                num3 = num1 + num2;
          }
          else if (source == bSub)
          {
                num3 = num1 - num2;
          }
          else if (source == bMul)
          {
                num3 = num1 * num2;
```

```
            }
            else if (source == bDiv)
            {
                  num3 = num1 / num2;
            }
            else if (source == bClear)
            {
                  tf1.setText("0.0");
                  tf2.setText("0.0");
                  tf3.setText("0.0");
            }
            else {} // do nothing
            // this also works
            // tf3.setText(new Double(num3).toString());
            tf3.setText("" + num3);
        }
    }

    private class MyCloseButtonHandler
                extends WindowAdapter  {
        public void windowClosing(WindowEvent we)
        {
                System.exit(0);
        }
    }

    public static void main(String args[])
    {
        SimpleCalculator sc = new SimpleCalculator();
        sc.startApp();
    }
}
```

To convert the application to Swing, we will begin by declaring a package for the application. Although the inclusion of a package declaration is advisable when preparing to create a `JAR` file, it is not strictly necessary at this stage if the `JAR` file is not immediately required.

Next, we will convert the AWT components and containers to their corresponding Swing components by adding a "`J`" prefix to each. For instance, we will use `JFrame`, `JPanel`, `JLabel`, `JTextField`, and `JButton` instead of their AWT counterparts. However, the layout managers will remain unchanged. For example, `GridLayout` should not have a "`J`" prefix as there is no equivalent Swing-specific layout manager.

To better illustrate the difference between AWT and Swing, a comparison can be made between the two versions of the same code. When running the AWT version on a Windows machine, the buttons and other UI elements appear less visually appealing compared to their Swing counterparts. Swing offers a more refined appearance, with customizable features such as background colors that provide a more modern look and feel. While the difference in appearance is subtle on macOS, the distinction is more noticeable on Windows.

Figure 11.3 Runtime Comparison between AWT and Swing Version

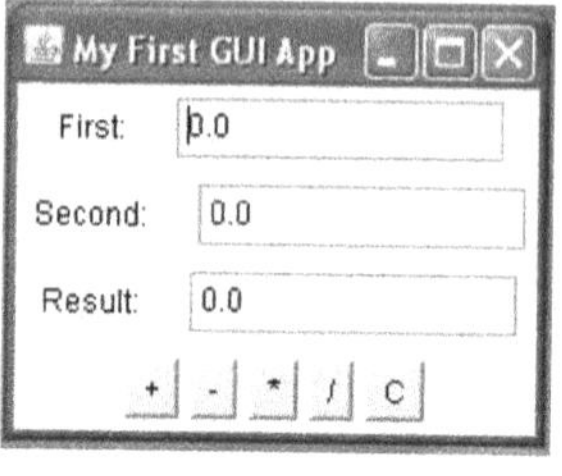

AWT Version

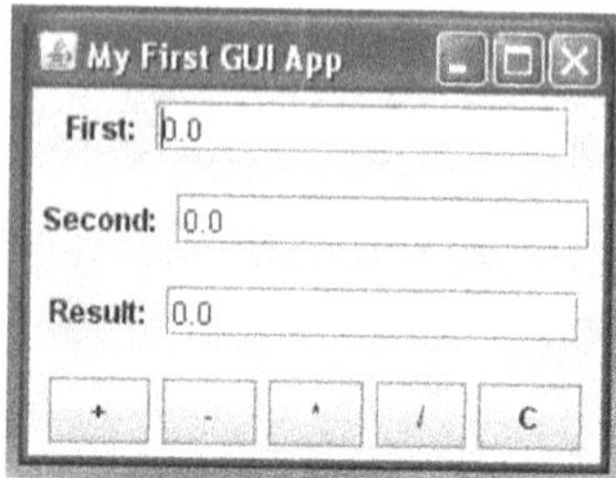

Swing Version

To initiate the conversion of our calculator application, we begin by modifying the component and container declarations, adding the "`J`" prefix where appropriate, both in the declarations and within the constructor, where components and containers are instantiated. This transforms them into Swing-based components. Importantly, no changes should be made to the layout managers, as Swing does not provide its own versions.

Regarding the close button handler, there is an option to retain the existing functionality, or it may be modified to use Swing's frame method for closing. Once the changes are saved, we can compile the code (using `javac SimpleCalculator.java`), and the application will now operate as a Swing-based component.

While the general appearance remains similar to the AWT version, the use of Swing provides enhanced functionality, and the UI elements—such as buttons—should now work as expected, including the "clear" and "close" buttons.

Here's the updated Swing version of the `SimpleCalculator` App:

```
// SimpleCalculator.java - updated Swing Version
import java.awt.*;
import java.awt.event.*;
import javax.swing.*;

public class SimpleCalculator implements ActionListener
{
    // Containers
    private JFrame f;
    private JPanel p1, p2, p3, p4;

    // Components
    private JLabel l1, l2, l3;
    private JTextField tf1, tf2, tf3;
    private JButton bAdd, bSub, bMul, bDiv, bClear;

    public SimpleCalculator()
    {
        // container
        f = new JFrame("My First GUI App");

        p1 = new JPanel();
        p2 = new JPanel();
        p3 = new JPanel();
        p4 = new JPanel();

        // components
        l1 = new JLabel("First: ");
        l2 = new JLabel("Second: ");
        l3 = new JLabel("Result: ");

        tf1 = new JTextField("0.0", 15);
```

```
        tf2 = new JTextField("0.0", 15);
        tf3 = new JTextField("0.0", 15);

        bAdd = new JButton("+");
        bSub = new JButton("-");
        bMul = new JButton("*");
        bDiv = new JButton("/");
        bClear = new JButton("C");
    }

    public void startApp()
    {
        // use the default layout manager for
        // all panels, use the FlowLayout.

        p1.add(l1);
        p1.add(tf1);

        p2.add(l2);
        p2.add(tf2);

        p3.add(l3);
        p3.add(tf3);

        p4.add(bAdd);
        p4.add(bSub);
        p4.add(bMul);
        p4.add(bDiv);
        p4.add(bClear);

        // change the layout manager for our Frame f;
        // use GridLayout(4, 1);
```

```
        f.setLayout(new GridLayout(4, 1));

        f.add(p1);
        f.add(p2);
        f.add(p3);
        f.add(p4);

        f.pack();
        f.setVisible(true);

        // register your event handler to
        // your UI (event source)
        bAdd.addActionListener(this);
        bSub.addActionListener(this);
        bMul.addActionListener(this);
        bDiv.addActionListener(this);
        bClear.addActionListener(this);

        f.addWindowListener(new MyCloseButtonHandler());
    }

    // event handler
    public void actionPerformed(ActionEvent e)
    {
        Object source = e.getSource();
        double num1 = 0.0, num2 = 0.0, num3 = 0.0;

        if (tf1.getText() != null
            && tf2.getText() != null)  {
            num1 = Double.parseDouble(tf1.getText());
            num2 = Double.parseDouble(tf2.getText());
```

```

                if (source == bAdd)
                {
                        num3 = num1 + num2;
                }
                else if (source == bSub)
                {
                        num3 = num1 - num2;
                }
                else if (source == bMul)
                {
                        num3 = num1 * num2;
                }
                else if (source == bDiv)
                {
                        num3 = num1 / num2;
                }
                else if (source == bClear)
                {
                        tf1.setText("0.0");
                        tf2.setText("0.0");
                        tf3.setText("0.0");
                }
                else {}         // do nothing
                // this also works
                // tf3.setText(new Double(num3).toString());
                tf3.setText("" + num3);
        }
    }

    private class MyCloseButtonHandler
```

```
                extends WindowAdapter  {
        public void windowClosing(WindowEvent we)
        {
                System.exit(0);
        }
    }

    public static void main(String args[])
    {
        SimpleCalculator sc = new SimpleCalculator();
        sc.startApp();
    }
}
```

11.3 Common Swing Components

11.3.1 `JCheckBox`: `RGBCheckBox.java`

The `JCheckBox` class in Java is a part of the Swing package, which provides a way to create checkboxes in a graphical user interface. These checkboxes can be either selected or deselected by the user. Unlike radio buttons, multiple checkboxes can be selected at the same time.

In the example `RGBCheckBox.java` given below, we will create a simple graphical interface with three checkboxes—Red, Green, and Blue—using Swing components. We will also add a button labeled "`Click Me!`" for further interaction.

Code Breakdown

1. Imports: The program starts by importing `java.awt.*` and `javax.swing.*`, which provide the necessary classes for creating the GUI components and handling the layout, the `java.awt` package is required for the layout managers.

```
import java.awt.*;
import javax.swing.*;
```

2. Class declaration: The class `RGBCheckBox` is declared as a public class. Inside this class, we define private fields for the `JFrame`, `JButton`, and `JCheckBox` components that will make up the UI.

```
public class RGBCheckBox {
   private JFrame f;
   private JButton b;
   private JCheckBox red, green, blue;
```

3. Constructor: The constructor `RGBCheckBox()` initializes the `JFrame` and `JButton` objects and creates three checkboxes: `red`, `green`, and `blue`. These components will be part of the user interface.

```
public RGBCheckBox() {
   f = new JFrame("JCheckBox Example");
   b = new JButton("Click Me!");
   red = new JCheckBox(" Red ");
   green = new JCheckBox(" Green ");
   blue = new JCheckBox(" Blue ");
}
```

4. StartApp method: The `startApp()` method configures the layout of the frame, which uses a `GridLayout` with 4 rows and 1 column to display the checkboxes and the button vertically. The `setSelected(true)` method is used to pre-select the Red and Blue checkboxes.

The `setLayout(new GridLayout(4, 1))` ensures that the components are arranged in four rows.

The `f.add()` method adds the components to the frame in the desired order.

`f.setBounds(100, 100, 300, 200)` specifies the position (100,100) of the top-left corner of the window and its size (300x200).

The `f.setVisible(true)` makes the window visible to the user.

```
public void startApp() {
    f.setLayout(new GridLayout(4, 1));
    red.setSelected(true);
    blue.setSelected(true);
    f.add(red);
    f.add(green);
    f.add(blue);
    f.add(b);
    f.setBounds(100, 100, 300, 200);
    f.setVisible(true);
}
```

5. Main method: The main method serves as the entry point for the program. It creates an instance of the `RGBCheckBox` class and calls the `startApp()` method to display the UI.

```
public static void main(String args[]) {
    RGBCheckBox rgb = new RGBCheckBox();
    rgb.startApp();
}
```

How the Code Works

When the program runs, a window appears with three checkboxes: Red, Green, and Blue. The Red and Blue checkboxes are pre-selected. Users can check or uncheck any combination of these boxes.

A button labeled "`Click Me!`" is also displayed, though it currently has no functionality implemented. Future extensions could involve adding event listeners to the button and checkboxes to perform actions when clicked.

Figure 11.4 *`RGBCheckBox`* **Runtime**

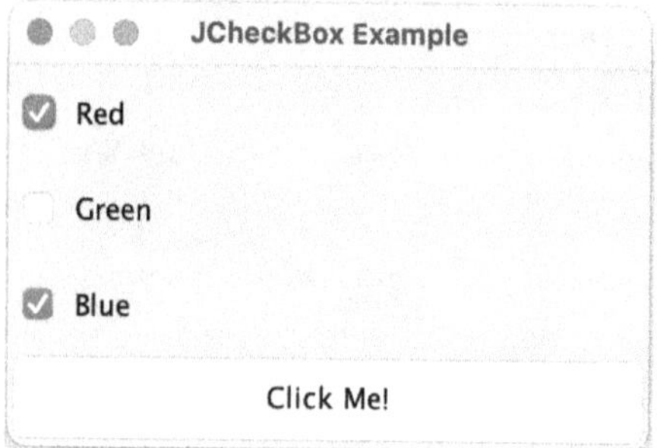

Key points

- `JCheckBox` is part of the Swing framework, and it inherits many methods from its superclass, allowing for extensive customization.

- The `setBounds()` method is used to both position and size the window, offering more control over its layout.

This simple example demonstrates how to create a basic GUI with multiple checkboxes using Swing components in Java. You can further extend this by adding event handling to make the UI interactive.

Here's the complete source code:

```
// RGBCheckBox.java

import java.awt.*;
import javax.swing.*;

public class RGBCheckBox
{
    private JFrame f;
    private JButton b;
    private JCheckBox red, green, blue;

    public RGBCheckBox()
    {
        f = new JFrame("JCheckBox Example");
        b = new JButton("Click Me!");
        red = new JCheckBox(" Red ");
        green = new JCheckBox(" Green ");
        blue = new JCheckBox(" Blue ");
    }

    public void startApp()
    {
        f.setLayout(new GridLayout(4, 1));
        red.setSelected(true);
```

```
            blue.setSelected(true);
            f.add(red);
            f.add(green);
            f.add(blue);
            f.add(b);
            f.setBounds(100, 100, 300, 200);
            f.setVisible(true);
        }

        public static void main(String args[])
        {
            RGBCheckBox rgb = new RGBCheckBox();
            rgb.startApp();
        }
    }
```

Instructions for Running

1. Save this code to a file named `RGBCheckBox.java`.
2. Compile the code using `javac RGBCheckBox.java`.
3. Run the program using `java RGBCheckBox`.

When you run it, a window will appear with three checkboxes (Red, Green, and Blue), with Red and Blue pre-selected, along with a "`Click Me!`" button.

11.3.2 JRadioButton: RGBRadioButton.java

In this explanation, we will break down how to create a basic Java Swing application using `JRadioButton` components, showing the difference between radio buttons and checkboxes.

Key Difference Between Radio Buttons and Checkboxes

- **Radio buttons**: Allow the user to select only one option at a time. When you select a radio button, any previously selected button in the same group is automatically deselected.
- **Checkboxes (`JCheckBox`)**: Allow the user to select multiple options simultaneously. You can select or deselect multiple checkboxes independently.

Code Breakdown

1. Class and import statements: We start by importing the necessary packages:

```
// For GridLayout and other layout managers
import java.awt.*;
// For JFrame, JButton, JRadioButton
import javax.swing.*;
```

2. Creating the main class: In this case, the class is called `RGBRadioButton`:

```
public class RGBRadioButton {
    // Declare the components:
    // JFrame, JButton, and JRadioButton
    private JFrame f;
    private JButton b;
    private JRadioButton red, green, blue;
    // Constructor to initialize the components
    public RGBRadioButton() {
        f = new JFrame("RGBJRadioButton Example");
        red = new JRadioButton(" Red ");  // Red
```

```
            green = new JRadioButton(" Green ");// Green
            blue = new JRadioButton(" Blue ");  // Blue
            b = new JButton("Click Me!"); // Button
        }
    }
```

3. The startApp method: This method sets up the layout and adds the components (radio buttons, buttons, etc.) to the frame:

```
public void startApp() {
    // Set a GridLayout with 4 rows and 1 column for
    // arranging components vertically
    f.setLayout(new GridLayout(4, 1));

    // By default, the Red radio button will
    // be selected at startup
    red.setSelected(true);

    // Add the radio buttons to the frame
    // so they appear in the UI
    f.add(red);
    f.add(green);


    f.add(blue);

    // Group the radio buttons together to ensure only
    // one can be selected at a time
    ButtonGroup colorGroup = new ButtonGroup();
    // add the red, green and blue
```

```
        // buttons to the group
        colorGroup.add(red);
        colorGroup.add(green);
        colorGroup.add(blue);

        // Add the "Click Me!" button to the frame
        f.add(b);

        // Set the size and location of the frame
        f.setBounds(200, 200, 300, 200);

        // Make the frame visible to the user
        f.setVisible(true);
    }
```

4. Main method: This method is the entry point of the program. It creates an instance of the `RGBRadioButton` class and calls the `startApp()` method to launch the application.

```
    public static void main(String args[]) {
        RGBRadioButton rgb = new RGBRadioButton();
        rgb.startApp();  // Start the application
    }
```

Key Concepts

- **`JRadioButton`**: Represents a radio button in Swing. Only one button in a group of radio buttons can be selected at a time. If one is selected, the other is automatically deselected.

- `ButtonGroup`: Groups the radio buttons together. Without this, each radio button would function independently, meaning multiple selections could be made.
- `GridLayout`: This layout manager arranges components in a grid of rows and columns. In this case, we use a 4x1 grid to display three radio buttons and a button vertically.
- `setSelected(true)`: This method pre-selects the Red radio button when the application starts.
- `setBounds()`: Defines the frame's position and size on the screen. In this case, the frame starts at (200, 200) with a width of 300 and a height of 200 pixels.

Grouping Radio Buttons

To make sure that only one radio button can be selected at a time, the buttons are grouped using a `ButtonGroup`. When the buttons are added to the group, it ensures mutual exclusivity:

```
ButtonGroup colorGroup = new ButtonGroup();
// add the red, green and blue buttons to the group
colorGroup.add(red);
colorGroup.add(green);
colorGroup.add(blue);
```

Visual Representation and Interaction

When you run the program, you'll see:

- Three radio buttons labeled Red, Green, and Blue
- A button labeled "Click Me!"
- The Red radio button is pre-selected by default

Clicking on any other radio button will automatically deselect the current selection, ensuring only one option is active at any given time.

Figure 11.5 `RGBRadioButton` **Runtime**

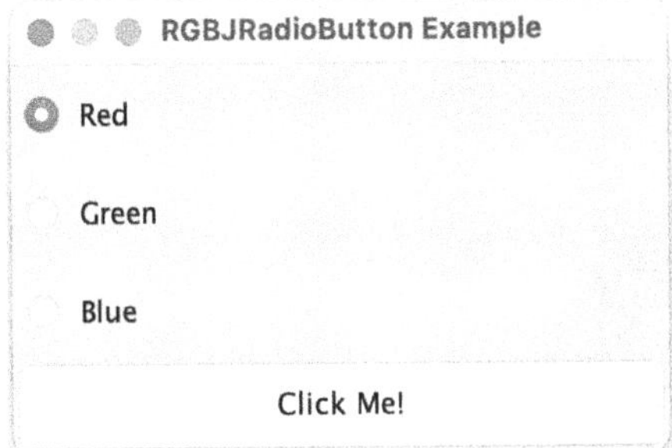

Exit Behavior

In Swing, even when the close button is clicked on the window, the application may not fully terminate unless a specific code is added to handle the window closure. The default behavior is that the window will hide but not exit unless specified otherwise.

Here's the full Java source code with a file named: `RGBRadioButton.java`

```
// RGBRadioButton.java

import java.awt.*;
import javax.swing.*;

public class RGBRadioButton {
    private JFrame f;
    private JButton b;
    private JRadioButton red, green, blue;
```

```

    public RGBRadioButton() {
        f = new JFrame("RGBJRadioButton Example");
        red = new JRadioButton(" Red ");
        green = new JRadioButton(" Green ");
        blue = new JRadioButton(" Blue ");
        b = new JButton("Click Me!");
    }

    public void startApp() {
        f.setLayout(new GridLayout(4, 1));
        red.setSelected(true);
        f.add(red);
        f.add(green);
        f.add(blue);

        ButtonGroup colorGroup = new ButtonGroup();
        colorGroup.add(red);
        colorGroup.add(green);
        colorGroup.add(blue);
        f.add(b);
        f.setBounds(200, 200, 300, 200);
        f.setVisible(true);
    }

    public static void main(String args[]) {
        RGBRadioButton rgb = new RGBRadioButton();
        rgb.startApp();
    }
}
```

11.3.3 JComboBox: JComboBoxExample.java

A JComboBox in Java is a Swing component that allows users to select an item from a drop-down list. It can either be a non-editable list where users pick from predefined values or an editable one where users can also type their own input. This component combines the functionality of a button and a text field, making it a versatile input method for users.

In this example, we'll walk through how to create a basic JComboBox, populate it with values, and implement event handling so that when a user selects an option, it updates a label to reflect the chosen value.

Components Used in the Code

- **JFrame**: A window or container that holds all the components of the application.
- **JButton**: A clickable button that triggers an event.
- **JComboBox**: The drop-down list that contains options for users to select from.
- **JLabel**: A label to display messages or results based on user actions.
- **String** Array: An array containing a list of colors, which will be displayed in the JComboBox.

Code Breakdown

1. **Declaring the class**

The class JComboBoxExample starts by declaring private variables to represent the JFrame, JButton, JComboBox, and JLabel, which will be used throughout the application.

```
private JFrame f;

private JButton b;
```

```
private JComboBox<String> cb;
private JLabel label;
private String colorList[];
```

- `f` represents the frame or window.
- `b` is the button labeled "Show".
- `cb` is the JComboBox containing color options.
- `label` will display the selected color.
- `colorList[]` holds the predefined color values for the JComboBox.

2. Constructor setup

The constructor initializes all the components of the GUI (Graphical User Interface). It creates the frame, button, `JComboBox`, and label, while also assigning positions using the `setBounds()` method, which defines their location and size within the frame.

```
public JComboBoxExample() {
    f = new JFrame("JComboBox List Example");
    label = new JLabel();
    colorList = new String[] {"Red", "Green",
                  "Blue", "Yellow", "Purple"};
    cb = new JComboBox<>(colorList);


    cb.setBounds(50, 100, 90, 20);
    label.setHorizontalAlignment(JLabel.CENTER);
    label.setSize(400, 100);

    b = new JButton("Show");
    b.setBounds(200, 100, 75, 20);
}
```

Here's what happens:

- A new `JFrame` is created with the title "JComboBox List Example".
- The `JLabel` is aligned to the center, and its size is set.
- The `JComboBox` is populated with the values from the `colorList` array.
- The `JButton` labeled "Show" is created.

3. Starting the application

The `startApp()` method is responsible for adding the components to the frame and making the frame visible.

```
public void startApp() {
    f.setLayout(null);      // No layout manager is used,
                            // so positions are set manually
                            // with setBounds.
    f.add(cb);              // Add JComboBox to the frame.
    f.add(label);           // Add label to the frame.
    f.add(b);               // Add button to the frame.

    f.setSize(350, 350);    // Set the size of the frame.
    f.setVisible(true);     // Make the frame visible.
}
```

The method also disables the default layout manager by calling `f.setLayout(null)`, allowing manual positioning of components.

4. Event handling for the button

To make the button functional, an `ActionListener` is added. When the button is clicked, the selected item from the `JComboBox` is retrieved and displayed in the `JLabel`.

```
b.addActionListener(new ActionListener() {
    public void actionPerformed(ActionEvent ae) {
        String str = "My Color List: "
            + cb.getItemAt(cb.getSelectedIndex());
        label.setText(str);  // Update the label with
                             // the selected color.
    }
});
```

- The `ActionListener` listens for a button click.
- When the button is clicked, the selected value from the `JComboBox` is retrieved using `cb.getSelectedIndex()` and is set in the label.
- The label text is updated with the selected color name.

5. Main method

The main method creates an instance of the `JComboBoxExample` class and starts the application by calling `startApp()`.

```
public static void main(String args[]) {
    JComboBoxExample jcbe = new JComboBoxExample();
    jcbe.startApp();
}
```

Output and Behavior

When the application is run:

- A window (JFrame) appears with a `JComboBox` containing the color options.
- The user can select a color from the drop-down.

- Upon clicking the "Show" button, the selected color is displayed on the label.

Figure 11.6 `JComboBoxExample` **Runtime 1**

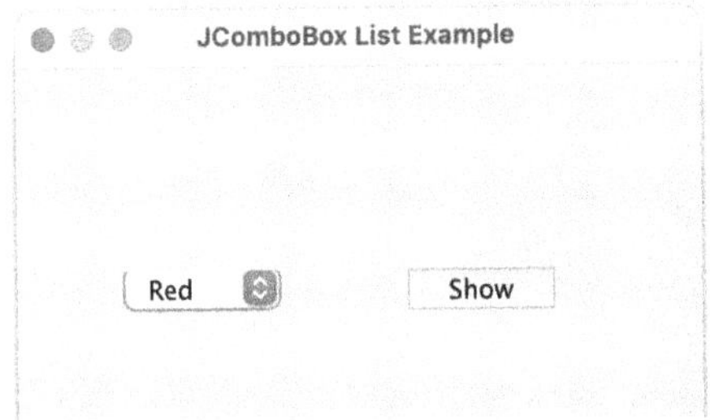

Figure 11.7 `JComboBoxExample` **Runtime 2**

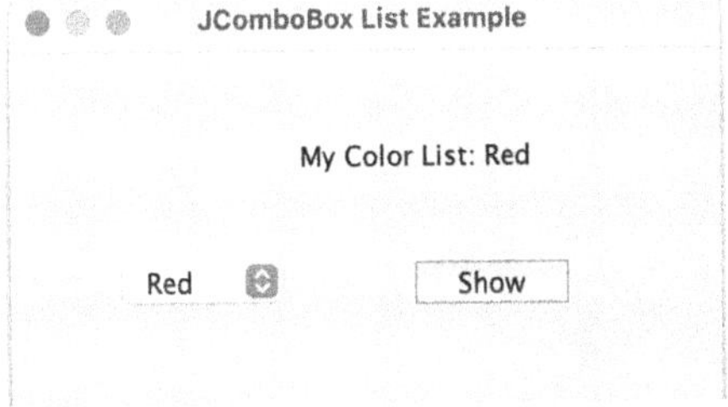

Here is the complete code for your convenience. Save it as `JComboBoxExample.java`

```
// JComboBoxExample.java
import javax.swing.*;
import java.awt.event.*;

public class JComboBoxExample {
   private JFrame f;
   private JButton b;
   private JComboBox<String> cb;
```

```
    private JLabel label;
    private String[] colorList;

    public JComboBoxExample() {
       f = new JFrame("JComboBox List Example");
       label = new JLabel();
       colorList = new String[] {"Red", "Green",
            "Blue", "Yellow", "Purple"};
       cb = new JComboBox<>(colorList);
       cb.setBounds(50, 100, 90, 20);
       label.setHorizontalAlignment(JLabel.CENTER);
       label.setSize(400, 100);
       b = new JButton("Show");
       b.setBounds(200, 100, 75, 20);
    }

   public void startApp() {
     f.setLayout(null);
     f.add(cb);
     f.add(label);
     f.add(b);
     f.setSize(350, 350);
     f.setVisible(true);
     b.addActionListener(new ActionListener() {
     public void actionPerformed(ActionEvent ae) {
        String str = "My Color List: "
          + cb.getItemAt(cb.getSelectedIndex());
          label.setText(str);
          }});
     }
```

```

    public static void main(String args[]) {
        JComboBoxExample jcbe = new JComboBoxExample();
        jcbe.startApp();
    }
}
```

11.3.4 JList: RGBJList.java

In Java, the `JList` component is used to display a list of items from which users can select one or more elements. With a `JList`, you have the flexibility to enable single or multiple selections, allowing users to choose several items at once.

To create a `JList`, it is necessary to have a "list model," which serves as a container for the elements you want to display in the `JList`. Think of the list model as the data holder, while the `JList` acts as the visual interface that shows the items.

In the provided sample code, a class named `RGBJList` is created. This class declares a frame (`JFrame`) and a label (`JLabel`) to display a list of color options. First, the frame is instantiated and set to use a `FlowLayout`, which helps arrange components from left to right, similar to text in a paragraph.

In the `startApp()` method, a `DefaultListModel` is created to hold the list data. This model is of type `String`, meaning it will store a collection of string values representing the color names. These values are added to the model using the `addElement()` method, which adds items such as "Red," "Green," and "Blue."

Once the list model is populated with these colors, a new `JList` is created and linked to the model. This connection ensures that the `JList` (`colorList` in the code) will display the items from the list model. By default, the first item in the list (index 0, "Red") is pre-selected using the `setSelectedIndex(0)` method.

Next, a `JLabel` is used to display some text ("Colors:"). Both the label and the `JList` are added to the frame, and the frame's size and position are set using `setBounds(200, 200, 300, 200)`, which specifies the frame's location and size in pixels.

Finally, the frame is made visible with the `setVisible(true)` method, allowing the application window to be displayed to the user.

When you run the code, you will see a list of color options ("Red," "Green," and "Blue"). Initially, "Red" is selected because of the `setSelectedIndex(0)` call. Users can click on any of the colors to make a selection. They can also make multiple selections by holding down the Shift or Command (Mac) / Control (Windows) keys while clicking on additional colors.

Here's the sample code saved as: **`RGBJList.java`**

```
// RGBJList.java
import java.awt.*;
import javax.swing.*;

    public class RGBJList {
     private JFrame f;
     private JLabel label;

     public RGBJList() {
```

```
        f = new JFrame("JList Color Example");
    }

    public void startApp() {
        f.setLayout(new FlowLayout());
        DefaultListModel<String> listModel =
            new DefaultListModel<>();
        listModel.addElement("Red");
        listModel.addElement("Green");
        listModel.addElement("Blue");

        JList<String> colorList = new JList<>();
        colorList.setModel(listModel);
        colorList.setSelectedIndex(0);
        label = new JLabel();
        label.setText("Colors: ");
        f.add(label);
        f.add(colorList);
        f.setBounds(200, 200, 300, 200);
        f.setVisible(true);
    }

    public static void main(String args[]) {
        RGBJList rgb = new RGBJList();
        rgb.startApp();
    }
}
```

This code creates a simple graphical interface with a `JList` displaying color names and enables interaction through single and multiple selections.

Figures 11.8 and 11.9 show a sample runtime.

Figure 11.8 `RGBJList` **Runtime with a single chosen option**

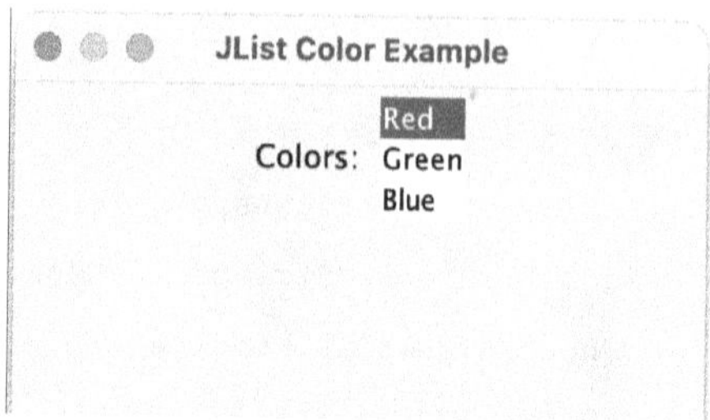

Figure 11.9 `RGBJList` **Runtime with multiple chosen options**

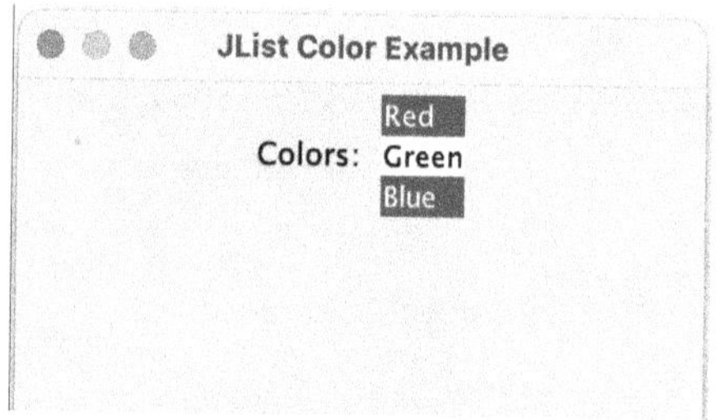

11.3.5 `JLabel: CounterApp.java`

Using `Label` in the `java.awt` package is too plain; the Swing version of `JLabel` provides a more exciting use of labels.

The following example breaks down the working of a `CounterApp` that is a basic counter application using Swing components such as `JLabel`, `JButton`, and event handling through `ActionListener`.

Figure 11.10 shows what it is supposed to look like.

Figure 11.10 `CounterApp` **Runtime**

This Java Swing application creates a counter that increments each time the "Click Me!" button is pressed. The counter value is displayed using three `JLabel` components representing the hundreds, tens, and ones places. The app leverages event handling to detect button clicks and updates the counter accordingly.

Explanation of Key Components

1. **Imports and packages**:
   - `java.awt.*`: This package is included to handle the layout of the components using classes like GridLayout and FlowLayout.
   - `java.awt.event.*`: This is needed to manage user interactions like button clicks, by handling ActionListener and ActionEvent.
   - `javax.swing.*`: Swing components like JFrame, JLabel, and JButton are used to build the graphical user interface (GUI).

2. **Class definition**: The main class, `CounterApp`, implements `ActionListener`, which allows the app to respond to user actions, such as button clicks.

3. **Instance variables:**
    - `JFrame f`: The main window of the application.
    - `Panel pan`: A container to hold the labels (counter digits).
    - `JLabel l1, l2, l3`: Labels to display the hundreds, tens, and ones digits of the counter.
    - `JButton button`: A button that increments the counter when clicked.
    - `int counter`: A static variable to store the current counter value.

4. **Constructor**: The constructor `CounterApp()` initializes the frame, panel, labels, and button. The labels are set with the initial value "0" and styled using the `Font` class (bold, serif, size 100). The background of the panel is set to black, while the button is initialized with the text "Click Me!".

5. **startApp method**: This method sets up the layout, adds components to the frame, and makes the window visible:
    - Panel layout: A grid layout with 1 row and 3 columns is used for the panel to align the labels horizontally.
    - Label configuration: Labels 1 and 2 are set to have a white foreground, and Label 3 (representing the ones place) has a black foreground with an opaque background for better visibility.
    - Frame configuration: The panel is added to the center of the frame, and the button is positioned at the bottom (south) of the frame using `BorderLayout.SOUTH`.
    - Event registration: The button is linked to an event handler using `button.addActionListener(this)`, which connects

the button click event to the `actionPerformed()` method.

6. **`actionPerformed` method**: When the button is clicked, this method is triggered:
   - The current counter value is retrieved by concatenating the values from the three labels and converting the result into an integer.
   - The counter is incremented by one (`counter++`), and the new value is split into hundreds, tens, and ones digits.
   - The text of each label is updated using `setText()`, converting the integer digits back to strings.

7. **Main method**: The `main` method creates an instance of the `CounterApp` class and starts the application by calling `startApp()`.

Key Concepts

- **JLabel customization:** The code demonstrates how to style JLabel components by changing fonts, colors, and opacity using methods like setFont(), setForeground(), and setOpaque().
- **Layout management:** The panel uses a grid layout to arrange the labels horizontally, while the frame uses a border layout to position components in specific regions (center for the panel and south for the button).
- **Event handling:** The application listens for user interactions (button clicks) using the ActionListener interface, making the app responsive to user input.

Running the Application

When the program runs, it displays a window with three large digits, all initialized to 0, and a button labeled "Click Me!" Every time the button is clicked, the counter increments, updating the labels to reflect the new counter value (with hundreds, tens, and ones positions).

In summary, this Java Swing application showcases a simple yet interactive counter using `JLabel` for display, `JButton` for interaction, and event handling to make the button functional. The use of Swing's flexible components allows for a highly customizable interface.

Here's the complete code: **`CounterApp.java`**

```
// CounterApp.java
import java.awt.*;
import java.awt.event.*;
import javax.swing.*;

public class CounterApp implements ActionListener {
  private JFrame f;
  private Panel pan;
  private JLabel l1, l2, l3;
  private JButton button;
  private static int counter = 0;

  public CounterApp() {
    f = new JFrame("My Counter App");
    pan = new Panel();
    pan.setBackground(Color.BLACK);
    l1 = new JLabel("0");
```

```
    l1.setFont(new Font("Serif", Font.BOLD, 100));
    l2 = new JLabel("0");
    l2.setFont(new Font("Serif", Font.BOLD, 100));
    l3 = new JLabel("0");
    l3.setFont(new Font("Serif", Font.BOLD, 100));
    button = new JButton("Click Me!");
  }

  public void startApp() {
    l3.setOpaque(true);
    l1.setForeground(Color.WHITE);
    l2.setForeground(Color.WHITE);
    l3.setForeground(Color.BLACK);
    pan.setLayout(new GridLayout(1, 3));
    pan.add(l1);
    pan.add(l2);
    pan.add(l3);
    f.add(pan, BorderLayout.CENTER);
    f.add(button, BorderLayout.SOUTH);
    f.pack();
    f.setVisible(true);
    f.setDefaultCloseOperation(JFrame.EXIT_ON_CLOSE);
    // register your event handler
    button.addActionListener(this);


  }

  public void actionPerformed(ActionEvent ae) {
    counter = Integer.parseInt(l1.getText() +
                l2.getText() + l3.getText());
    counter++;
```

```
    int hund = counter / 100;
    int temp = counter % 100;
    int tens = temp / 10;
    int ones = temp % 10;
    l1.setText("" + hund);
    l2.setText("" + tens);
    l3.setText("" + ones);
  }

  public static void main(String args[]) {
    CounterApp counterApp = new CounterApp();
    counterApp.startApp();
  }
}
```

11.4 Packaging Your Application to a Java ARchive (JAR) File

11.4.1 JAR Files Overview

JAR stands for *Java Archive* and is a file format used to distribute Java applications or libraries in a single file. Similar to ZIP files, JAR files bundle multiple files together, but they are designed specifically for Java-related files, such as `.class` files, resources, and metadata. JAR files make it easy to distribute Java programs and their associated files across different platforms. There are two common types of JAR files:

- **Regular JAR file:** Bundles Java classes and resources, but doesn't specify a main class.
- **Executable JAR file:** Includes a `MANIFEST.MF` file that declares the main class, making the JAR file directly executable.

11.4.2 Step-By-Step Guide for Creating JAR Files

Setting up a basic project directory

Before creating a JAR file, you need to organize your Java project files. Suppose you are working on a Swing Desktop App with the following directory structure:

```
SwingDesktopApp/
|
|——src/        // Contains your source code (.java files)
|   |
|   |——demo/
|   |
|   └——SimpleCalculator.java
|
|——classes/    // Will store compiled .class files
|
└——dist/       // Will store the final JAR files
```

Key directories:

`src/`: Contains your source code (`.java` files)

`classes/`: This is where the compiled `.class` files will be stored.

`dist/`: This is where the `JAR` file will be stored after packaging.

Adding the package declaration to your Java file

When converting your application into a `JAR` file, it's crucial to ensure that the Java classes are part of a package. For example, if your `SimpleCalculator.java` file is inside the `src/demo/` folder, your source code should start with the following package declaration:

```
// should be placed on top of your Java Source Code.

package demo;
```

The package declaration should reflect the folder structure of your project. The folder that contains your .java file (in this case, `demo/`) should be the package name.

Compiling the Java source code (command line)

To compile the source code, open a terminal and navigate to the `src/` directory of your project. The following command will compile the `SimpleCalculator.java` file:

```
javac -d ../classes demo/SimpleCalculator.java
```

Explanation:

- `javac`: Is the Java compiler command
- `-d ../classes`: Specifies the output directory for the compiled .class files (in this case, the classes/ folder)
- `demo/SimpleCalculator.java`: The path to the Java source file to be compiled

Once compiled, the classes/ folder will look like this:

```
classes/
└── demo/
    └── SimpleCalculator.class
```

Creating a regular JAR file (command line)

Now that you have your compiled `.class` files, you can create a regular `JAR` file (without a manifest file). From the classes/ directory, run the following command:

```
jar cvf MyApp.jar demo/
```

Explanation:

- `jar`: Command to create or manage JAR files
- `cvf`: Options used for creating the JAR file:
  - `c`: Create a new JAR
  - `v`: Verbose output (optional, for showing details during creation)

    - `f`: Specifies the name of the JAR file
- `MyApp.jar`: The name of the resulting JAR file
- `demo/`: The folder containing the compiled .class files to include in the JAR

This command will create a file named `MyApp.jar` in the `classes/` directory, which contains the contents of the `demo/` folder.

Running a regular JAR file (command line)

A regular JAR file doesn't have a manifest file, so to run it, you need to specify the classpath and the fully qualified name of the main class. The command would look like this:

```
java -cp MyApp.jar demo.SimpleCalculator
```

Explanation:

- `java`: The Java runtime command
- `-cp MyApp.jar`: Specifies the JAR file as the classpath
- `demo.SimpleCalculator`: The fully qualified name of the main class (including the package name)

Creating an executable JAR file (command line)

An executable JAR file contains a `MANIFEST.MF` file that specifies which class should be executed when the JAR is run. Here's how to create one:

Step 1: Create a manifest file

A manifest file contains metadata about the JAR file, including the main class to be executed. Create a file named manifest.txt in the classes/ directory with the following content:

```
Main-Class: demo.SimpleCalculator
```

Important: Make sure there is a blank line at the end of the manifest file. This is required by the manifest format.

Step 2: Create the executable JAR

Use the following command to create the JAR file with the manifest included:

```
jar cvfm MyAppExecutable.jar manifest.txt demo/
```

Explanation:

- `cvfm`: The m option is used to specify the manifest file.
- `MyAppExecutable.jar`: The name of the resulting executable JAR file.
- `manifest.txt`: The path to the manifest file.
- `demo/`: The folder containing the compiled `.class` files.

Step 3: Running the executable JAR file

After creating the executable JAR, you can run it simply by using the java -jar command:

```
java -jar MyAppExecutable.jar
```

Since the `MANIFEST.MF` file contains the `Main-Class` entry; you don't need to specify the class name or the classpath explicitly.

Creating JAR files using the NetBeans IDE

If you're using an IDE like `NetBeans`, the process of creating JAR files is simplified.

Step 1: Create a new Java project in NetBeans

1. Open NetBeans and create a new Java project (`File > New Project`).
2. Choose `Java with Ant` as the project type and give your project a name like `SwingDesktopAppProject`.
3. Ensure that "`Create Main Class`" is unchecked during project setup, as you'll manually add the class later.

Step 2: Add source code and package

1. In the `Projects` window, right-click on `Source Packages`, and choose `New > Java Class`.
2. Create a package named `demo` and a class named `SimpleCalculator` (same as before).
3. Add the Java code for your calculator to the `SimpleCalculator.java` file, including the package declaration.

Step 3: Run and test your application

1. Right-click on `SimpleCalculator.java` in the `Projects` tab and select `Run File`.
2. NetBeans will compile and execute your application. You should see your calculator in action.

Step 4: Create the executable JAR in NetBeans

1. Add the `Main-Class` declaration: Go to the `Files` tab in NetBeans. Open the `manifest.mf` file located under the project folder.
2. Add the following entry to the `manifest.mf` file:

```
Main-Class: demo.SimpleCalculator
```

Ensure there's a blank line after this entry.

3. **Build the project:** Right-click on your project in the `Projects` tab and select `Clean and Build`. This will compile your project and package it into a `JAR` file.
4. **Locate the JAR:** The executable `JAR` file will be located in the `dist/` folder of your project directory.

Step 5: Run the executable JAR

You can run the `JAR` file by navigating to the dist/ folder in a terminal and using the command:

```
java -jar SwingDesktopAppProject.jar
```

By following the steps above, you can create both regular and executable `JAR` files using either the command line or an IDE like `NetBeans`. The critical difference between the two types of `JAR` files lies in the presence of a `MANIFEST.MF` file with a `Main-Class` declaration, which allows you to run the `JAR` file without specifying the classpath or main class explicitly.

Chapter Summary

- Java AWT and Swing are both used for building GUIs, but AWT relies on native OS components, making it platform-dependent, while Swing provides lightweight, platform-independent components that look consistent across platforms. Swing also offers more features, such as additional components and support for a customizable look and feel, and follows the MVC design pattern, unlike AWT.
- This lesson focuses on converting an AWT-based calculator application to a Swing-based one by importing the javax.swing package. The transformation involves replacing AWT components, such as Frame and Button, with their Swing counterparts, like JFrame and JButton, while retaining the same layout managers and event handling. Although both versions function similarly, Swing offers enhanced visual appeal and platform-independent UI features, which are particularly noticeable on Windows systems.
- A JAR (Java ARchive) file is used to bundle Java applications or libraries into a single file, making distribution easier. There are two main types: regular JARs, which bundle class files but don't specify a main class, and executable JARs, which include a manifest file with a `Main-Class` declaration for direct execution. You can create a JAR file by compiling your Java code, organizing it into directories, and using the `jar` command. Alternatively, an IDE like NetBeans simplifies this process, automatically packaging your project and creating the necessary manifest file for executable JARs.

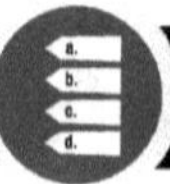

Multiple Choice Questions

1. **What package should you import if you want to use the Swing Framework?**

   a. `javax.swing` package
   b. `java.swing` package
   c. `java.awt` package
   d. `java.awt.swing` package

2. **Which of the following is NOT a valid GUI / Event package that we can import?**

   a. `javax.swing.*;`
   b. `java.swing.*;`
   c. `java.awt.*;`
   d. `java.awt.event.*;`

3. **You are planning to create an executable `JAR` file, and your test class is under `classes\myFolder\MyTestClass.class`. Your current folder (your present working directory) is in `classes\`. Which of the following commands will allow you to create an executable `JAR` file assuming you have the correct manifest file named `manif.mf` in the same folder (`classes`)?**

   a. `jar cvfm MyJar.jar manif.mf`
   b. `jar cvf MyJar.jar manif.mf`
   c. `jar cvf MyJar.jar *`
   d. `jar cvfm MyJar.jar manif.mf *`

4. **You have created a non-executable** `JAR` **file called** `MyJar.jar`**. The test class is located in** `classes\myFolder\MyTestClass.class`**. You created the** `JAR` **file while inside the** `classes` **folder. Which of the following commands is the correct way of running the** `JAR` **file?**
   - **a.** `java -jar myFolder.MyTestClass`
   - **b.** `jar MyJar.jar myFolder.MyTestClass`
   - **c.** `java -cp MyJar.jar myFolder.MyTestClass`
   - **d.** `java -jar MyJar.jar`

5. **What does JAR stand for?**
   - **a.** Java Archive Resource
   - **b.** Java Application Resource
   - **c.** Java Automatic Resource
   - **d.** Java ARchive

6. **What do you call the special file that can contain information about the files packaged in a JAR file?**
   - **a.** class file
   - **b.** manifest file
   - **c.** Java Source File
   - **d.** Executable File

7. **Which of the following IS NOT TRUE about Swing?**
   a. Unlike AWT, Java Swing provides a set of platform-independent and lightweight components.
   b. The Swing Package is built on top of AWT (Abstract Window Toolkit).
   c. You can find the Frame, the Button, and the Label classes under the `javax.swing` package.
   d. Java Swing Package is a part of Java Foundation Classes (JFC) that is used to create window-based applications.

8. **Which Swing component is an implementation of a radio button that can select or deselect an item, which is used with a `ButtonGroup` object to create a group of buttons in which only one item at a time can be selected?**
   a. `JComboBox`
   b. `JCheckbox`
   c. `JRadioButton`
   d. `JList`

9. **What Swing component displays a list of objects and allows the user to select one or more items? This will also require you to have a `ListModel` that will maintain the contents of the list that the user can choose from.**
   a. `JList`
   b. `JRadioButton`
   c. `JComboBox`
   d. `JCheckbox`

10. **Which of the following IS NOT TRUE about Swing Components?**
    a. Swing components are platform-independent.
    b. Swing provides fewer components compared to AWT
    c. Swing Components are lightweight components.
    d. Swing supports pluggable look and feel components.

Multiple Choice Answer Key

1 – a	2 – b	3 – d	4 – c	5 – d
6 – b	7 – c	8 – c	9 – a	10 – b

Coding Tasks

- Revisit the Coding Task from **Chapter 10: My Color Calculator (with Event Handler)** and convert your **AWT Color Calculator** to a **Swing Calculator App**.
- Create an executable `JAR` file for your Color Calculator so that when you double-click your `JAR` file, it will launch your Color Calculator.

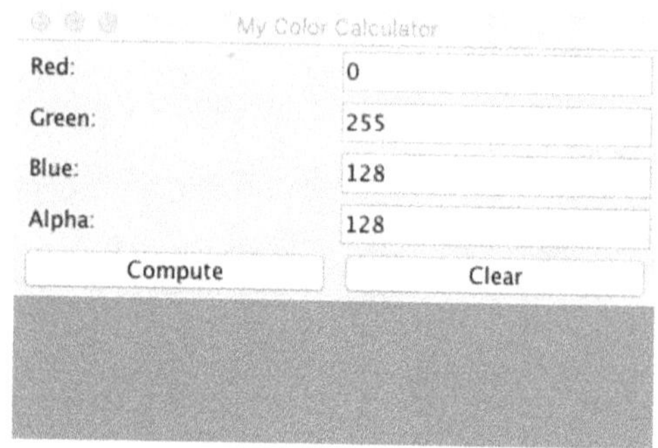

www.ingramcontent.com/pod-product-compliance
Lightning Source LLC
Chambersburg PA
CBHW070743220326
41598CB00026B/3726
9781636516561